BRITANNICA

Mathematics in Context

Michigan Mathematics
Inservice Project

Reallotment

D1301456

⊕ **Britannica**
ENCYCLOPÆDIA BRITANNICA EDUCATIONAL CORPORATION

Mathematics in Context is a comprehensive curriculum for the middle grades. It was developed in collaboration with the Wisconsin Center for Education Research, School of Education, University of Wisconsin–Madison and the Freudenthal Institute at the University of Utrecht, The Netherlands, with the support of National Science Foundation Grant No. 9054928.

National Science Foundation

Opinions expressed are those of the authors
and not necessarily those of the Foundation

ISBN 0-7826-1516-3
1 2 3 4 5 WK 02 01 00 99 98

The *Mathematics in Context* Development Team

Mathematics in Context is a comprehensive curriculum for the middle grades. The National Science Foundation funded the National Center for Research in Mathematical Sciences Education at the University of Wisconsin–Madison to develop and field-test the materials from 1991 through 1996. The Freudenthal Institute at the University of Utrecht in The Netherlands, as a subcontractor, collaborated with the University of Wisconsin–Madison on the development of the curriculum.

The initial version of *Reallotment* was developed by Koeno Gravemeijer. It was adapted for use in American schools by Barbara Clarke and Margaret A. Pligge.

National Center for Research in Mathematical Sciences Education Staff

Thomas A. Romberg
Director

Joan Daniels Pedro
Assistant to the Director

Gail Burrill
Coordinator
Field Test Materials

Margaret R. Meyer
Coordinator
Pilot Test Materials

Mary Ann Fix
Editorial Coordinator

Sherian Foster
Editorial Coordinator

James A. Middleton
Pilot Test Coordinator

Margaret A. Pligge
First Edition Coordinator

Project Staff

Jonathan Brendefur
Laura J. Brinker
James Browne
Jack Burrill
Rose Byrd
Peter Christiansen
Barbara Clarke
Doug Clarke
Beth R. Cole

Fae Dremock
Jasmina Milinkovic
Mary C. Shafer
Julia A. Shew
Kay Schultz
Aaron N. Simon
Marvin Smith
Stephanie Z. Smith
Mary S. Spence
Kathleen A. Steele

Freudenthal Institute Staff

Jan de Lange
Director

Els Feijs
Coordinator

Martin van Reeuwijk
Coordinator

Project Staff

Mieke Abels
Nina Boswinkel
Frans van Galen
Koeno Gravemeijer
Marja van den Heuvel-Panhuizen
Jan Auke de Jong
Vincent Jonker
Ronald Keijzer

Martin Kindt
Jansie Niehaus
Nanda Querelle
Anton Roodhardt
Leen Streefland
Adri Treffers
Monica Wijers
Astrid de Wild

Acknowledgments

Several school districts used and evaluated one or more versions of the materials: Ames Community School District, Ames, Iowa; Parkway School District, Chesterfield, Missouri; Stoughton Area School District, Stoughton, Wisconsin; Madison Metropolitan School District, Madison, Wisconsin; Milwaukee Public Schools, Milwaukee, Wisconsin; and Dodgeville School District, Dodgeville, Wisconsin. Two sites were involved in staff development as well as formative evaluation of materials: Culver City, California, and Memphis, Tennessee. Two sites were developed through partnership with Encyclopædia Britannica Educational Corporation: Miami, Florida, and Puerto Rico. University Partnerships were developed with mathematics educators who worked with preservice teachers to familiarize them with the curriculum and to obtain their advice on the curriculum materials. The materials were also used at several other schools throughout the United States.

We at Encyclopædia Britannica Educational Corporation extend our thanks to all who had a part in making this program a success. Some of the participants instrumental in the program's development are as follows:

Allapattah Middle School
Miami, Florida
Nemtalla (Nikolai) Barakat

Ames Middle School
Ames, Iowa
Kathleen Coe
Judd Freeman
Gary W. Schnieder
Ronald H. Stromen
Lyn Terrill

Bellerive Elementary
Creve Coeur, Missouri
Judy Hetterscheidt
Donna Lohman
Gary Alan Nunn
Jakke Tchang

Brookline Public Schools
Brookline, Massachusetts
Rhonda K. Weinstein
Deborah Winkler

Cass Middle School
Milwaukee, Wisconsin
Tami Molenda
Kyle F. Witty

Central Middle School
Waukesha, Wisconsin
Nancy Reese

Craigmont Middle School
Memphis, Tennessee
Sharon G. Ritz
Mardest K. VanHooks

Crestwood Elementary
Madison, Wisconsin
Diane Hein
John Kalson

Culver City Middle School
Culver City, California
Marilyn Culbertson
Joel Evans
Joy Ellen Kitzmiller
Patricia R. O'Connor
Myrna Ann Perks, Ph.D.
David H. Sanchez
John Tobias
Kelley Wilcox

Cutler Ridge Middle School
Miami, Florida
Lorraine A. Valladares

Dodgeville Middle School
Dodgeville, Wisconsin
Jacqueline A. Kamps
Carol Wolf

Edwards Elementary
Ames, Iowa
Diana Schmidt

Fox Prairie Elementary
Stoughton, Wisconsin
Tony Hjelle

Grahamwood Elementary
Memphis, Tennessee
M. Lynn McGoff
Alberta Sullivan

Henry M. Flagler Elementary
Miami, Florida
Frances R. Harmon

Horning Middle School
Waukesha, Wisconsin
Connie J. Marose
Thomas F. Clark

Huegel Elementary
Madison, Wisconsin
Nancy Brill
Teri Hedges
Carol Murphy

Hutchison Middle School
Memphis, Tennessee
Maria M. Burke
Vicki Fisher
Nancy D. Robinson

Idlewild Elementary
Memphis, Tennessee
Linda Eller

Jefferson Elementary
Santa Ana, California
Lydia Romero-Cruz

Jefferson Middle School
Madison, Wisconsin
Jane A. Beebe
Catherine Buege
Linda Grimmer
John Grueneberg
Nancy Howard
Annette Porter
Stephen H. Sprague
Dan Takkunen
Michael J. Vena

Jesus Sanabria Cruz School
Yabucoa, Puerto Rico
Andreíta Santiago Serrano

John Muir Elementary School
Madison, Wisconsin
Julie D'Onofrio
Jane M. Allen-Jauch
Kent Wells

Kegonsa Elementary
Stoughton, Wisconsin
Mary Buchholz
Louisa Havlik
Joan Olsen
Dominic Weisse

Linwood Howe Elementary
Culver City, California
Sandra Checel
Ellen Thireos

Mitchell Elementary
Ames, Iowa
Henry Gray
Matt Ludwig

New School of Northern Virginia
Fairfax, Virginia
Denise Jones

Northwood Elementary
Ames, Iowa
Eleanor M. Thomas

Orchard Ridge Elementary
Madison, Wisconsin
Mary Paquette
Carrie Valentine

Parkway West Middle School
Chesterfield, Missouri
Elissa Aiken
Ann Brenner
Gail R. Smith

Ridgeway Elementary
Ridgeway, Wisconsin
Lois Powell
Florence M. Wasley

Roosevelt Elementary
Ames, Iowa
Linda A. Carver

Roosevelt Middle
Milwaukee, Wisconsin
Sandra Simmons

Ross Elementary
Creve Coeur, Missouri
Annette Isselhard
Sheldon B. Korklan
Victoria Linn
Kathy Stamer

St. Joseph's School
Dodgeville, Wisconsin
Rita Van Dyck
Sharon Wimer

St. Maarten Academy
St. Peters, St. Maarten, NA
Shareed Hussain

Sarah Scott Middle School
Milwaukee, Wisconsin
Kevin Haddon

Sawyer Elementary
Ames, Iowa
Karen Bush Hoiberg

Sennett Middle School
Madison, Wisconsin
Brenda Abitz
Lois Bell
Shawn M. Jacobs

Sholes Middle School
Milwaukee, Wisconsin
Chris Gardner
Ken Haddon

Stephens Elementary
Madison, Wisconsin
Katherine Hogan
Shirley M. Steinbach
Kathleen H. Vegter

Stoughton Middle School
Stoughton, Wisconsin
Sally Bertelson
Polly Goepfert
Jacqueline M. Harris
Penny Vodak

Toki Middle School
Madison, Wisconsin
Gail J. Anderson
Vicky Grice
Mary M. Ihlenfeldt
Steve Jernegan
Jim Leidel
Theresa Loehr
Maryann Stephenson
Barbara Takkunen
Carol Welsch

Trowbridge Elementary
Milwaukee, Wisconsin
Jacqueline A. Nowak

W. R. Thomas Middle School
Miami, Florida
Michael Paloger

Wooddale Elementary Middle School
Memphis, Tennessee
Velma Quinn Hodges
Jacqueline Marie Hunt

Yahara Elementary
Stoughton, Wisconsin
Mary Bennett
Kevin Wright

Site Coordinators

Mary L. Delagardelle—Ames Community Schools, Ames, Iowa

Dr. Hector Hirigoyen—Miami, Florida

Audrey Jackson—Parkway School District, Chesterfield, Missouri

Jorge M. López—Puerto Rico

Susan Militello—Memphis, Tennessee

Carol Pudlin—Culver City, California

Reviewers and Consultants

Michael N. Bleicher
Professor of Mathematics
University of Wisconsin–Madison
Madison, WI

Diane J. Briars
Mathematics Specialist
Pittsburgh Public Schools
Pittsburgh, PA

Donald Chambers
Director of Dissemination
University of Wisconsin–Madison
Madison, WI

Don W. Collins
Assistant Professor of Mathematics Education
Western Kentucky University
Bowling Green, KY

Joan Elder
Mathematics Consultant
Los Angeles Unified School District
Los Angeles, CA

Elizabeth Fennema
Professor of Curriculum and Instruction
University of Wisconsin-Madison
Madison, WI

Nancy N. Gates
University of Memphis
Memphis, TN

Jane Donnelly Gawronski
Superintendent
Escondido Union High School
Escondido, CA

M. Elizabeth Graue
Assistant Professor of Curriculum and Instruction
University of Wisconsin–Madison
Madison, WI

Jodean E. Grunow
Consultant
Wisconsin Department of Public Instruction
Madison, WI

John G. Harvey
Professor of Mathematics and Curriculum & Instruction
University of Wisconsin–Madison
Madison, WI

Simon Hellerstein
Professor of Mathematics
University of Wisconsin–Madison
Madison, WI

Elaine J. Hutchinson
Senior Lecturer
University of Wisconsin–Stevens Point
Stevens Point, WI

Richard A. Johnson
Professor of Statistics
University of Wisconsin–Madison
Madison, WI

James J. Kaput
Professor of Mathematics
University of Massachusetts–Dartmouth
Dartmouth, MA

Richard Lehrer
Professor of Educational Psychology
University of Wisconsin–Madison
Madison, WI

Richard Lesh
Professor of Mathematics
University of Massachusetts–Dartmouth
Dartmouth, MA

Mary M. Lindquist
Callaway Professor of Mathematics Education
Columbus College
Columbus, GA

Baudilio (Bob) Mora
Coordinator of Mathematics & Instructional Technology
Carrollton-Farmers Branch Independent School District
Carrollton, TX

Paul Trafton
Professor of Mathematics
University of Northern Iowa
Cedar Falls, IA

Norman L. Webb
Research Scientist
University of Wisconsin–Madison
Madison, WI

Paul H. Williams
Professor of Plant Pathology
University of Wisconsin–Madison
Madison, WI

Linda Dager Wilson
Assistant Professor
University of Delaware
Newark, DE

Robert L. Wilson
Professor of Mathematics
University of Wisconsin–Madison
Madison, WI

TABLE OF CONTENTS

BRITANNICA

Mathematics in Context

Dear Teacher,

Welcome! *Mathematics in Context* is designed to reflect the National Council of Teachers of Mathematics Standards for School Mathematics and to ground mathematical content in a variety of real-world contexts. Rather than relying on you to explain and demonstrate generalized definitions, rules, or algorithms, students investigate questions directly related to a particular context and construct mathematical understanding and meaning from that context.

The curriculum encompasses 10 units per grade level. *Reallotment* is designed to be the first unit in the geometry strand for grade 6/7, but it also lends itself to independent use—to introduce students to the relationship between three-dimensional shapes and their two-dimensional representations.

In addition to the Teacher Guide and Student Books, *Mathematics in Context* offers the following components that will inform and support your teaching:

- *Teacher Resource and Implementation Guide,* which provides an overview of the complete system, including program implementation, philosophy, and rationale

- *Number Tools,* Volumes 1 and 2, which are a series of blackline masters that serve as review sheets or practice pages involving number issues and basic skills

- *News in Numbers,* which is a set of additional activities that can be inserted between or within other units; it includes a number of measurement problems that require estimation.

Thank you for choosing *Mathematics in Context.* We wish you success and inspiration!

Sincerely,

The Mathematics in Context Development Team

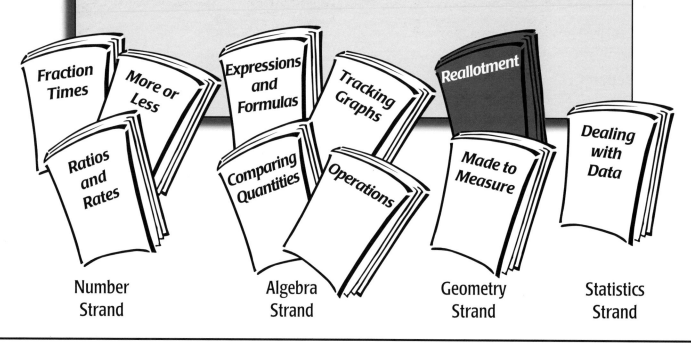

| Number Strand | Algebra Strand | Geometry Strand | Statistics Strand |

Overview

How to Use This Book

This unit is one of 40 for the middle grades. Each unit can be used independently; however, the 40 units are designed to make up a complete, connected curriculum (10 units per grade level). There is a Student Book and a Teacher Guide for each unit.

Each Teacher Guide comprises elements that assist the teacher in the presentation of concepts and in understanding the general direction of the unit and the program as a whole. Becoming familiar with this structure will make using the units easier.

Each Teacher Guide consists of six basic parts:

- Overview
- Student Materials and Teaching Notes
- Assessment Activities and Solutions

- Glossary
- Blackline Masters
- Try This! Solutions

Overview

Before beginning this unit, read the Overview in order to understand the purpose of the unit and to develop strategies for facilitating instruction. The Overview provides helpful information about the unit's focus, pacing, goals, and assessment, as well as explanations about how the unit fits with the rest of the *Mathematics in Context* curriculum.

Student Materials and Teaching Notes

This Teacher Guide contains all of the student pages (except the Try This! activities), each of which faces a page of solutions, samples of students' work, and hints and comments about how to facilitate instruction. Note: Solutions for the Try This! activities can be found at the back of the Teacher Guide.

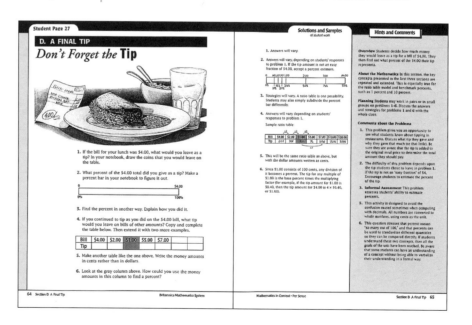

Each section within the unit begins with a two-page spread that describes the work students do, the goals of the section, new vocabulary, and materials needed, as well as providing information about the mathematics in the section and ideas for pacing, planning instruction, homework, and assessment.

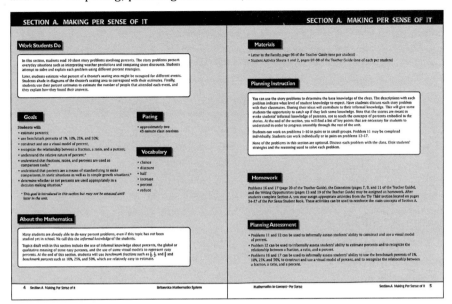

Assessment Activities and Solutions

Information about assessment can be found in several places in this Teacher Guide. General information about assessment is given in the Overview; informal assessment opportunities are identified on the teacher pages that face each student page; and the Assessment Activities section of this guide provides formal assessment opportunities.

Glossary

The Glossary defines all vocabulary words listed on the Section Opener pages. It includes mathematical terms that may be new to students, as well as words associated with the contexts introduced in the unit. (Note: The Student Book does not have a glossary. This allows students to construct their own definitions, based on their personal experiences with the unit activities.)

Blackline Masters

At the back of this Teacher Guide are blackline masters for photocopying. The blackline masters include a letter to families (to be sent home with students before beginning the unit), several student activity sheets, and assessment masters.

Try This! Solutions

Also included in the back of this Teacher Guide are the solutions to several Try This! activities—one related to each section of the unit—that can be used to reinforce the unit's main concepts. The Try This! activities are located in the back of the Student Book.

Unit Focus

The purpose of this unit is to broaden students' understanding of area by involving them in a wide range of real-life situations. This is in sharp contrast to the conventional introduction to area—using a formula such as *length times width* or counting squares. Students are encouraged to use a variety of strategies to solve problems. Several examples focus on finding the areas of irregular shapes. This is to emphasize that the concept of area is also relevant to shapes that are not rectangles, squares, parallelograms, triangles, or circles. Students also explore relationships between area, perimeter, surface area, and volume.

Mathematical Content

- estimating and computing the areas of figures and irregularly shaped figures
- generalizing formulas to estimate and compute the circumference and area of a circle
- understanding which measurement units and tools are appropriate in different situations
- understanding the structure and use of the metric system
- analyzing the relationships between the perimeter of a figure and its area
- identifying, describing, and classifying geometric figures, such as regular polygons
- exploring the concept of surface area
- creating and working with tessellations
- using strategies that involve transformations to estimate and compute the areas of figures
- using the concepts of area, perimeter, and volume to solve realistic problems

Prior Knowledge

This unit assumes that students have:

- performed whole number calculations,
- used a calculator,
- become familiar with basic geometric shapes (square, rectangle, quadrilateral, triangle, hexagon, circle, cylinder, and so on).

Planning and Preparation

Pacing: 20–23 days

Section	Work Students Do	Pacing*	Materials
A. The Size of Shapes	■ compare the sizes of shapes ■ study tessellations	4–5 days	■ Letter to the Family (one per student) ■ Student Activity Sheets 1–4 (one of each per student) ■ See page 5 for a complete list of the materials and quantities needed.
B. Areas	■ compare the areas of a variety of shapes ■ estimate and measure the areas of geometric figures by realloting sections to make new shapes	4–5 days	■ Student Activity Sheets 5–8 (one of each per student) ■ See page 27 for a complete list of the materials and quantities needed.
C. Area Patterns	■ transform a rectangle into a parallelogram in order to understand area relationships ■ calculate the areas of a variety of shapes using different strategies and units of measure that are not squares	4 days	■ Student Activity Sheets 9–11 (one of each per student) ■ See page 41 for a complete list of the materials and the quantities needed.
D. Measuring Area	■ study relationships between standard measuring units ■ calculate population densities	4 days	■ See page 67 for a complete list of the materials and the quantities needed.
E. Perimeter, Area, and Volume	■ explore perimeter, area, and volume relationships using a square, hexagon, circle, cylinder, and box.	4–5 days	■ Student Activity Sheets 4 and 12 (one of each per student) ■ See page 85 for a complete list of the materials and the quantities needed.

* One day is approximately equivalent to one 45-minute class session.

Preparation

In the *Teacher Resource and Implementation Guide* is an extensive description of the philosophy underlying both the content and the pedagogy of the *Mathematics in Context* curriculum. Suggestions for preparation are also given in the Hints and Comments columns of this Teacher Guide. You may want to consider the following:

• Work through the unit before teaching it. If possible, take on the role of the student and discuss your strategies with other teachers.

• Use the overhead projector for student demonstrations, particularly with overhead transparencies of the student activity sheets and any manipulatives used in the unit.

• Invite students to use drawings and examples to illustrate and clarify their answers.

• Allow students to work at different levels of sophistication. Some students may need concrete materials, while others can work at a more abstract level.

• Provide opportunities and support for students to share their strategies, which often differ. This allows students to take part in class discussions and introduces them to alternative ways to think about the mathematics in the unit.

• In some cases, it may be necessary to read the problems to students or to pair students to facilitate their understanding of the printed materials.

• A list of the materials needed for this unit is in the chart on page xiii.

• Try to follow the recommended pacing chart on page xiii. You can easily spend more time on this unit than the number of class periods indicated. Bear in mind, however, that many of the topics introduced in this unit will be revisited and covered more thoroughly in other *Mathematics in Context* units.

Resources

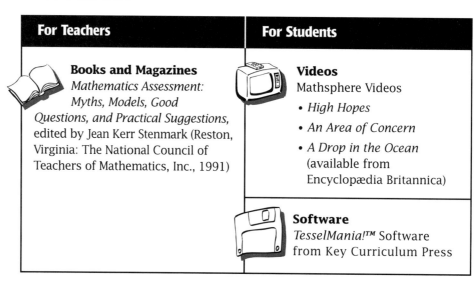

For Teachers	For Students
Books and Magazines *Mathematics Assessment: Myths, Models, Good Questions, and Practical Suggestions,* edited by Jean Kerr Stenmark (Reston, Virginia: The National Council of Teachers of Mathematics, Inc., 1991)	**Videos** Mathsphere Videos • *High Hopes* • *An Area of Concern* • *A Drop in the Ocean* (available from Encyclopædia Britannica)
	Software *TesselMania!™* Software from Key Curriculum Press

Assessment

Planning Assessment

In keeping with the NCTM Assessment Standards, valid assessment should be based on evidence drawn from several sources. (See the full discussion of assessment philosophies in the *Teacher Resource and Implementation Guide*.) An assessment plan for this unit may draw from the following sources:

- Observations—look, listen, and record observable behavior.

- Interactive Responses—in a teacher-facilitated situation, note how students respond, clarify, revise, and extend their thinking.

- Products—look for the quality of thought evident in student projects, test answers, worksheet solutions, or writings.

These categories are not meant to be mutually exclusive. In fact, observation is a key part of assessing interactive responses and also key to understanding the end results of projects and writings.

Ongoing Assessment Opportunities

- **Problems within Sections**
  To evaluate ongoing progress, *Mathematics in Context* identifies informal assessment opportunities and the goals that these particular problems assess throughout the Teacher Guide. There are also indications as to what you might expect from your students.

- **Section Summary Questions**
  The summary questions at the end of each section are vehicles for informal assessment (see Teacher Guide pages 24, 38, 62, 64, 82, and 108).

End-of-Unit Assessment Opportunities

In the back of this Teacher Guide, there are two assessments that when combined will take approximately three to four 45-minute class sessions. For a more detailed description of these assessment activities, see the Assessment Overview (Teacher Guide pages 110 and 111).

You may also wish to design your own culminating project or let students create one that will tell you what they consider important in the unit. For more assessment ideas, refer to the charts on pages xvi and xvii.

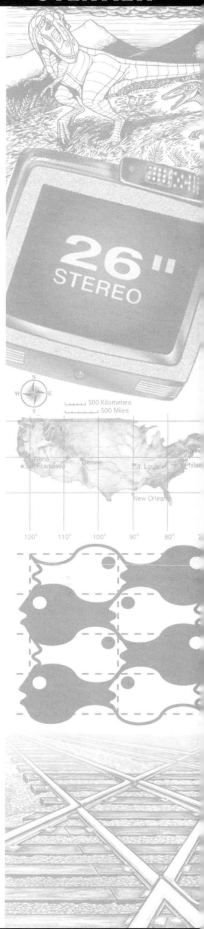

Goals and Assessment

In the *Mathematics in Context* curriculum, unit goals, categorized according to cognitive procedures, relate to the strand goals and the NCTM Curriculum and Evaluation Standards. Additional information about these goals is found in the *Teacher Resource and Implementation Guide.* The *Mathematics in Context* curriculum is designed to help students develop their abilities so that they can perform with understanding in each of the categories listed below. It is important to note that the attainment of goals in one category is not a prerequisite to attaining those in another category. In fact, students should progress simultaneously toward several goals in different categories.

	Goal	Ongoing Assessment Opportunities	End-of-Unit Assessment Opportunities
Conceptual and Procedural Knowledge	**1.** identify, describe, and classify geometric figures	**Section A** p. 14, #6	Rellotment Problems, pp. 138-141
	2. compare the areas of shapes using a variety of strategies and measuring units	**Section A** p. 14, #6 **Section B** p. 32, #9 **Section C** p. 48, #9 p. 52, #14	Sizing up Islands, p. 137 Rellotment Problems, pp. 138-141

	3. estimate and compute the areas of geometric figures	**Section A** p. 24, #13c
		Section B p. 32, #8, #10,
		                p. 34, #11
		Section C p. 52, #14
		                p. 64, #25
			Reallotment Problems, pp. 138-141

	4. create and work with tessellation patterns	**Section A** p. 22, #11 p. 24, #13 **Section C** p. 56, #18	Reallotment Problems, pp. 138-141

	Goal	Ongoing Assessment Opportunities	End-of-Unit Assessment Opportunities

Reasoning, Communicating, Thinking, and Making Connections	**5.** understand which units and tools are appropriate to estimate and measure area, perimeter, and volume	**Section C** p. 52, #14
		         p. 56, #18
		         p. 64, #25
		Section D p. 78, #15
		         p. 82, #18
		Section E p. 108, #29, #30
			Sizing up Islands, p. 137
			Reallotment Problems, pp. 138–141

	6. understand the structure and use of standard systems of measurement, both metric and English	**Section D** p. 72, #7
		         p. 82, #18
		Section E p. 100, #24
		         p. 104, #27b
		         p. 108, # 29, #30
			Reallotment Problems, pp. 138–141

	7. use the concepts of perimeter, area, and volume to solve realistic problems	**Section B** p. 34, #11
		Section C p. 52, #14
		         p. 56, #18
		Section D p. 76, #12
		         p. 78, #15
		Section E p. 106, #28
			Sizing up Islands, p. 137
			Reallotment Problems, pp. 138–141

	Goal	Ongoing Assessment Opportunities	End-of-Unit Assessment Opportunities
Modeling, Nonroutine Problem-Solving, Critically Analyzing, and Generalizing	**8.** represent and solve problems using geometric models	**Section B** p. 32, #8 **Section C** p. 54, #16 **Section E** p. 106, #28	Reallotment Problems, pp. 138–141
	9. analyze the effect a systematic change in dimension has on area, perimeter, and volume	**Section E** p. 88, #6 p. 108, #31	Reallotment Problems, pp. 138–141
	10. generalize formulas and procedures for determining the areas of rectangles, triangles, parallelograms, quadrilaterals, and circles	**Section C** p. 44, #4, #5 p. 50, #13 **Section E** p. 96, #19	Reallotment Problems, pp. 138–141

More about Assessment

Scoring and Analyzing Assessment Responses

Students may respond to assessment questions with various levels of mathematical sophistication and elaboration. Each student's response should be considered for the mathematics that it shows, and not judged on whether or not it includes an expected response. Responses to some of the assessment questions may be viewed as either correct or incorrect, but many answers will need flexible judgment by the teacher. Descriptive judgments related to specific goals and partial credit often provide more helpful feedback than percentage scores.

Openly communicate your expectations to all students, and report achievement and progress for each student relative to those expectations. When scoring students' responses try to think about how they are progressing toward the goals of the unit and the strand.

Student Portfolios

Generally, a portfolio is a collection of student-selected pieces that is representative of a student's work. A portfolio may include evaluative comments by you or by the student. See the *Teacher Resource and Implementation Guide* for more ideas on portfolio focus and use.

A comprehensive discussion about the contents, management, and evaluation of portfolios can be found in *Mathematics Assessment: Myths, Models, Good Questions, and Practical Suggestions*, pp. 35–48.

Student Self-Evaluation

Self-evaluation encourages students to reflect on their progress in learning mathematical concepts, their developing abilities to use mathematics, and their dispositions toward mathematics. The following examples illustrate ways to incorporate student self-evaluations as one component of your assessment plan.

- Ask students to comment, in writing, on each piece they have chosen for their portfolios and on the progress they see in the pieces overall.
- Give a writing assignment entitled "What I Know Now about [a math concept] and What I Think about It." This will give you information about each student's disposition toward mathematics as well as his or her knowledge.
- Interview individuals or small groups to elicit what they have learned, what they think is important, and why.

Suggestions for self-inventories can be found in *Mathematics Assessment: Myths, Models, Good Questions, and Practical Suggestions*, pp. 55–58.

Summary Discussion

Discuss specific lessons and activities in the unit—what the student learned from them and what the activities have in common. This can be done in whole-class discussions, small groups, or in personal interviews.

Connections across the *Mathematics in Context* Curriculum

Reallotment is the third unit in the geometry strand and the first geometry unit at this grade level. The map below shows the complete *Mathematics in Context* curriculum for grade 6/7. It indicates where the unit fits in the geometry strand, and where it fits in the overall picture.

A detailed description of the units, the strands, and the connections in the *Mathematics in Context* curriculum can be found in the *Teacher Resource and Implementation Guide.*

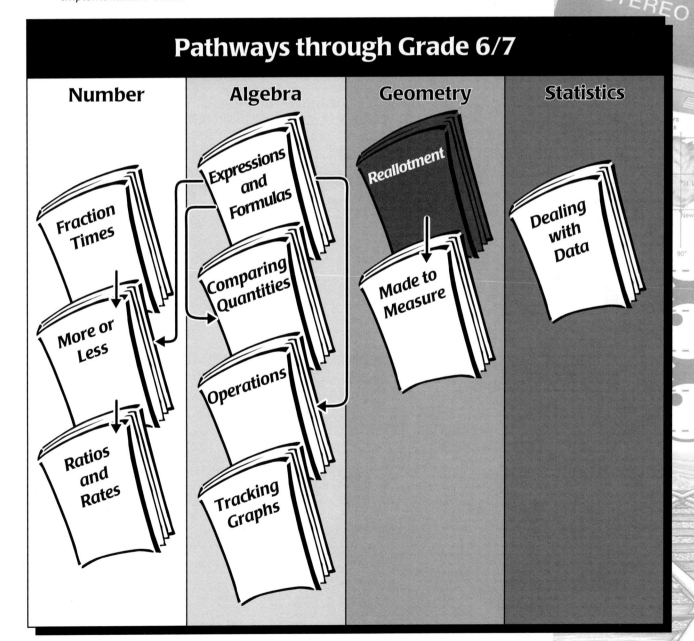

Pathways through Grade 6/7

Number	Algebra	Geometry	Statistics
Fraction Times	Expressions and Formulas	Reallotment	Dealing with Data
More or Less	Comparing Quantities	Made to Measure	
Ratios and Rates	Operations		
	Tracking Graphs		

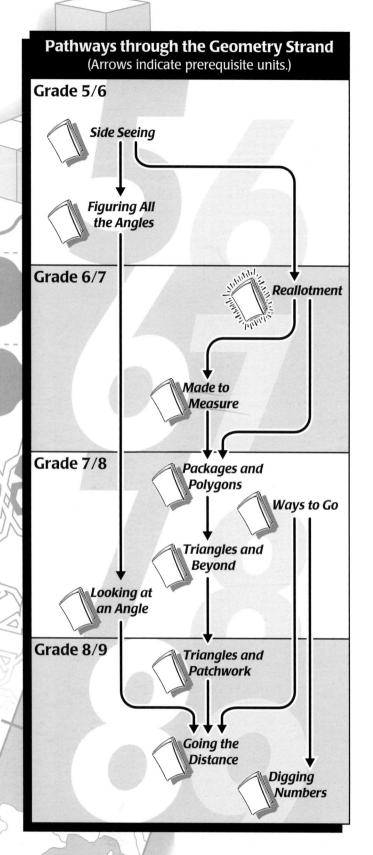

Pathways through the Geometry Strand
(Arrows indicate prerequisite units.)

Grade 5/6

Side Seeing

Figuring All the Angles

Grade 6/7

Reallotment

Made to Measure

Grade 7/8

Packages and Polygons

Ways to Go

Triangles and Beyond

Looking at an Angle

Grade 8/9

Triangles and Patchwork

Going the Distance

Digging Numbers

Connections within the Geometry Strand

On the left is a map of the geometry strand; this unit, *Reallotment,* is highlighted.

Reallotment is the third unit in the geometry strand and is preceded by the grade 5/6 units, *Side Seeing* and *Figuring All the Angles.*

Understanding shapes is one of the main substrands of the geometry strand. Although the contents of units within this substrand will be familiar to most teachers, the teaching philosophy is not traditional. In this unit, an informal approach to finding the area of a shape leads to a more formal notion of the reasoning behind area formulas. The process of reallotting parts of a shape in order to estimate its area helps students develop a strong, fundamental conception of area that will be useful in other geometry units. The main focus is on estimation, and computation of irregularly shaped figures. A variety of strategies are developed. Relationships between perimeter, area, and volume are also investigated in this unit.

Both the exploration of space and the more traditional geometry of forms and shapes are parts of the units *Packages and Polygons, Triangles and Beyond, Triangles and Patchwork,* and *Going the Distance.*

The Geometry Strand

Grade 5/6

Side Seeing
Exploring the relationship between three-dimensional shapes and drawings of them, seeing from different points of view, and building structures from drawings.

Figuring All the Angles
Estimating and measuring angles and investigating direction, vectors, and rectangular and polar coordinates.

Grade 7/8

Packages and Polygons
Recognizing geometric shapes in real objects and representations, constructing models, and investigating properties of regular and semi-regular polyhedra.

Looking at an Angle
Recognizing vision lines in two and three dimensions; identifying and drawing shadows and blind spots; identifying the isomorphism of vision lines, light rays, flight paths, and so forth; understanding the relationship between angles and the tangent ratio; and computing with the tangent ratio.

Ways to Go
Reading and interpreting different kinds of maps, comparing different types of distances, progressing from one two-dimensional model to another (from a diagram to a map to a photograph to a graph), and drawing graphs and networks. (*Ways to Go* is also in the statistics strand.)

Triangles and Beyond
Exploring the interrelationships of the sides and angles of triangles as well as the properties of parallel lines and quadrilaterals; constructing triangles; and using transformations to become familiar with the concepts of congruence and similarity.

Grade 6/7

Reallotment
Measuring regular and irregular areas; discovering links between area, perimeter, surface area, and volume; and using English and metric units.

Made to Measure
Measuring length (including circumference), volume, and surface area using metric units.

Grade 8/9

Triangles and Patchwork
Understanding similarity and using it to find unknown measurements for similar triangles and developing the concept of ratio through tessellation.

Going the Distance
Using the Pythagorean theorem to investigate distances, scales, and vectors and using slope, tangent, area, square root, and contour lines.

Digging Numbers
Using the properties of height, diameter, and radius to determine whether or not various irregular shapes are similar; predicting length using graphs and formulas; exploring the relationship between three-dimensional shapes and drawings of them; and using length-to-width ratios to classify various objects. (*Digging Numbers* is also in the statistics strand.)

Connections with Other *Mathematics in Context* Units

Understanding shapes is one of three substrands of the geometry strand. The exploration of space in the units in this strand and the more traditional geometry of forms and shapes come together in the units *Packages and Polygons, Triangles and Beyond, Triangles and Patchwork*, and *Going the Distance*.

The following mathematical topics included in this unit are revisited or further developed in other *Mathematics in Context* units.

Topic	Unit	Grade
measurement	*Made to Measure*	6/7
	*Insights into Data**	8/9
perimeter	*Made to Measure*	6/7
	*Ratios and Rates***	6/7
	Packages and Polygons	7/8
	*Cereal Numbers***	7/8
area	*Ratios and Rates***	6/7
	*Building Formulas****	7/8
	Packages and Polygons	7/8
	*Cereal Numbers***	7/8
	*Patterns and Figures****	8/9
	Triangles and Patchwork	8/9
	*Insights into Data**	8/9
shapes	*Ratios and Rates***	6/7
	*Building Formulas****	7/8
	Packages and Polygons	7/8
	*Cereal Numbers***	7/8
	*Patterns and Figures****	8/9
	Triangles and Beyond	7/8
	Triangles and Patchwork	8/9
transformations	*Triangles and Beyond*	7/8
	Triangles and Patchwork	8/9
volume	*Made to Measure*	6/7
	*Cereal Numbers***	7/8
	Going the Distance	8/9

  * This unit in the statistics strand also helps students make connections to ideas about geometry.
  ** These units in the number strand also help students make connections to ideas about geometry.
*** These units in the algebra strand also help students make connections to ideas about geomerty.

Student
Materials
and Teaching
Notes

Student Book
Table of Contents

Dear Student,

Welcome to the unit *Reallotment*.

In this unit you will study different shapes and the space covered by a variety of shapes.

You will figure out things such as how many people can stand in your classroom. How could you find out without packing people in the entire classroom?

You will also investigate the border around a shape and the amount of space inside a three-dimensional figure.

How can you make a shape like the one below that will cover a floor, leaving no open spaces?

In the end, you will have learned some important ideas about algebra, geometry, and arithmetic. We hope you enjoy the unit.

Sincerely,

The Mathematics in Context Development Team

Work Students Do

Students develop and use different strategies to compare the areas of different-sized shapes. For example, students place one shape on the top of another and look at the overlapping sections. They also use nonconventional units of measure, such as dot patterns, to compare the areas of shapes (they estimate or count the numbers of dots in two shapes). Students develop an understanding of the concept of reallotment—they discover that the area of a shape remains the same when reshaped by cutting and pasting.

Students use a square tile with a given price as a measuring unit to determine the sizes and prices of tiles of other shapes. The number of measuring units needed to cover a shape is called the *area* of a shape. In the last part of this section, students create tessellation patterns by beginning with a basic shape, cutting portions out of one part, and pasting them onto another part. This reinforces the idea of reallotting areas.

Goals

Students will:

- identify, describe, and classify geometric figures;

- compare the areas of shapes using a variety of strategies and measuring units;

- estimate and compute the area of geometric figures;

- create and work with tessellation patterns;

- understand which units and tools are appropriate to estimate and measure area, perimeter, and volume.*

  * *This goal is introduced in this section and is assessed later in the unit.*

Pacing

- approximately four or five 45-minute class sessions

Vocabulary

- area
- measuring unit
- tessellation

About the Mathematics

The concept of *area* —the number of measuring units needed to cover a shape—is implicitly introduced. Students use estimation and informally use the concepts of ratio and proportion to compare areas of different-sized shapes. The mathematical term *area* is not introduced until after students have experienced filling the interior of a two–dimensional shape. The square measuring unit is introduced as a mathematical convention and related to the relative cost of different-sized tiles.

Tessellations are patterns that fill a plane using congruent copies of a figure that do not overlap. For example, a tessellation can be made by covering a plane using regular triangles. Not all geometric figures can be used to make a tessellation, however. For example, regular pentagons or octagons do not tessellate (fill in the entire space without overlapping).

The strategies students use to compare the areas of shapes in this section are not only important in developing their understanding of area and their ability to determine area, but also lay a foundation that will help them better understand how formal area formulas are derived in Section C. These strategies generally deal with reallotting (reshaping) a shape. A shape can be seen as the sum of other shapes or as a portion of another shape. A shape can also be rearranged to form a different shape by cutting and pasting.

Materials

- Letter to the Family, page 124 of the Teacher Guide (one per student)
- Student Activity Sheets 1–4, pages 125–128 of the Teacher Guide (one of each per student)
- tracing paper, pages 7, 11, 13, 17, 19, and 25 of the Teacher Guide, optional (one sheet per student)
- grid paper, pages 7, 17, 19, 21, 23 and 25 of the Teacher Guide, (one sheet per student)
- scissors, pages 7, 9, 11, 13, 15, 17, and 23 of the Teacher Guide, optional (one pair per student)
- glue or tape, pages 7, 9, 15, and 23 of the Teacher Guide, optional (one bottle or roll per pair or group of students)
- rulers, page 13 of the Teacher Guide (one per student)
- drawing paper, page 13 of the Teacher Guide (one sheet per student)
- transparency, page 17 of the Teacher Guide, optional (one per class)
- compasses, page 17 of the Teacher Guide, optional (one per student)
- overhead projector, page 17 of the Teacher Guide, optional (one per class)
- colored pencils, page 21 of the Teacher Guide, optional (one box per student)

Planning Instruction

Organize a class discussion about students' solutions to problems 3, 4e, 4h, 4j, and 8. You might also discuss the names of some of the geometric shapes on page 4 of the Student Book (square, rectangle, triangle, and so on). You may wish to elaborate on problem 11 and have students create a tessellation design. They can then describe their creations and find the areas of their designs.

Students may work individually or in pairs on problems 12 and 13. They may work in pairs or in small groups on the remaining problems in this section.

There are no optional problems in this section.

Homework

Problems 6 (page 14 of the Teacher Guide), 8 (page 16 of the Teacher Guide), 9 (page 18 of the Teacher Guide), 10 (page 20 of the Teacher Guide), 11 (page 22 of the Teacher Guide), and 12 (page 24 of the Teacher Guide) can be assigned as homework. The Writing Opportunity (page 11 of the Teacher Guide), the Extensions (pages 17 and 21 of the Teacher Guide), and the Interdisciplinary Connection (page 19 of the Teacher Guide) can also be assigned as homework. After students complete Section A, you may assign appropriate activities from the Try This! section, located on pages 49–52 of the Student Book. The Try This! activities reinforce the key mathematical concepts introduced in this section.

Planning Assessment

- Problem 6 can be used to informally assess students' ability to identify, describe, and classify geometric figures; and to compare areas of shapes using a variety of strategies and measuring units.
- Problems 11 and 13 can be used to informally assess students' ability to create and work with tessellation patterns.
- Problem 13c can be used to informally assess students' ability to estimate and compute the areas of geometric figures.

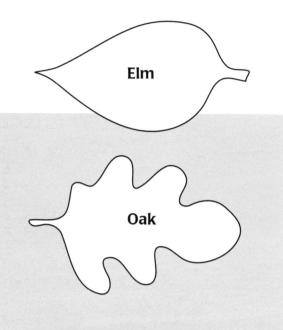

Elm

Oak

Leaves and Trees

Look at the two leaf outlines on the left. Suppose that one side of each leaf will be frosted with a thin layer of chocolate.

1. Which leaf will have more chocolate? Explain your reasoning.

The map on the left shows two forests separated by a river and a swamp. Meg and Tom are having a discussion about which forest is larger.

2. Use the figures below to decide which forest is larger. Then describe the method you used.

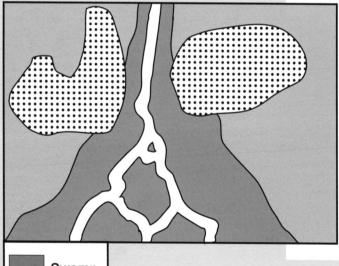

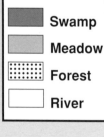

■	Swamp
▨	Meadow
▦	Forest
□	River

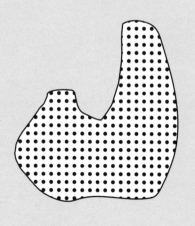

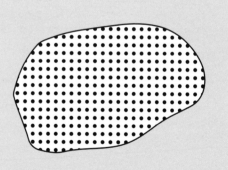

1. The oak leaf will have more chocolate frosting, but the leaves are very close in size. Students may trace one leaf and place it over the other leaf to see the non-overlapping areas. Some students may need to cut and paste their tracings.

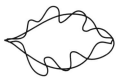

Students may also use grid paper to count the number of complete squares the leaves cover. They can combine leftover pieces to form new squares and then compare the number of squares each leaf covers.

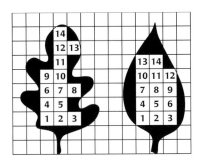

2. The forest on the right is slightly larger. Students may use the same tracing strategy mentioned for problem **1.** Some students may count the number of dots covered by each forest and compare them. Students may divide the forests into rectangles and calculate the number of dots in each, as shown below.

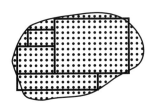

Students may also combine tracing and counting strategies and count only the dots in the non-overlapping areas, as shown below. The white spaces show the overlap between the two shapes.

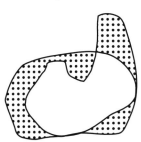

Materials tracing paper, optional (one sheet per pair or group of students), grid paper, optional, (one sheet per pair or group of students), scissors (one pair per student), glue or tape (one bottle or roll per pair or group of students), all optional

Overview Students informally explore the concept of area by developing their own methods to compare the sizes of different shapes. The vocabulary word *area* is not yet introduced.

About the Mathematics The context of the problems on this page encourage students to focus on the concept of area and should discourage them from confusing area with perimeter. Several strategies for comparing the areas of different shapes can be used:

• putting one shape on top of the other and then focusing on the overlap;

• cutting and pasting one shape onto another shape to see whether or not it can cover the other shape;

• counting the number of dots on each shape.

Problem **2** shows that squares and other figures, such as dots, can be used to compare the sizes of shapes. Each dot represents a part of the shape. This strategy informally introduces the concept of area: the number of measuring units needed to cover a shape.

Planning Encourage students to invent their own strategies using any of the materials listed above to compare the areas of the shapes. Students can work in pairs or in small groups on problems **1** and **2.** After students finish, review the various methods they used to solve the problems.

Comments about the Problems

1–2. Students' discussions while working in pairs or small groups may motivate them to find more accurate ways to compare the areas of shapes.

2. Observe as students work on this problem. If you notice that students count all the dots one by one, suggest that they count the number of dots in one row and use that information to estimate the number of dots in a larger section.

Tulip Fields

Here you see three fields of tulips. Use **Student Activity Sheet 1** to solve the following problem.

3. Which field has the most tulip plants? Justify your answer.

Field A   **Field B**

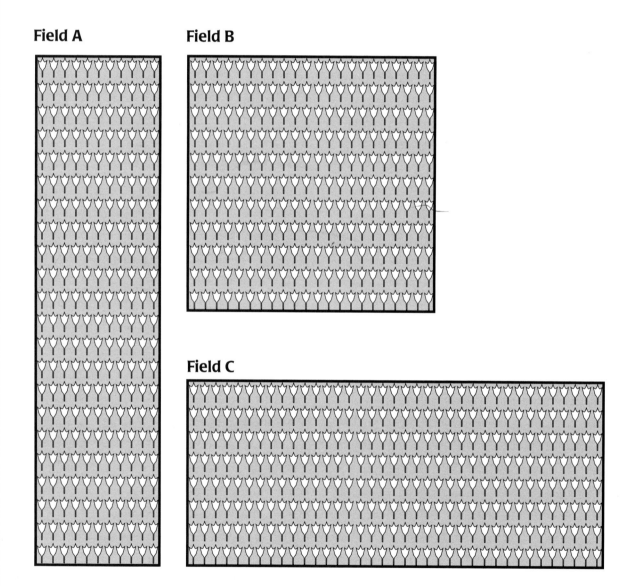

Field C

3. Field C. It has 296 tulips, while fields A and B each have 242. Students may multiply the number of tulips in one row by the number of tulips in one column.

Field A: 22 × 11 = 242 tulips

Field B: 11 × 22 = 242 tulips

Field C: 8 × 37 = 296 tulips.

Students may also trace and compare fields A and B as illustrated below.

Field A

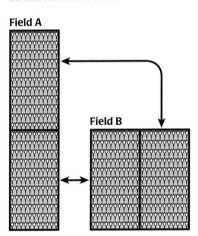

Field B

Materials Student Activity Sheet 1 (one per student); scissors, optional (one pair per student); glue or tape, optional (one bottle or roll per pair or group of students)

Overview Students compare the areas of three tulip fields and determine, using a variety of strategies, which field has the most tulips.

About the Mathematics Some students may compare the areas of the tulip fields by cutting and rearranging the shape of one field and pasting it onto the shape of another field. This strategy is based on a property of area: when a shape is subdivided, cut, and/or rearranged, the original area remains intact.

Other students may use the strategy of counting and/or estimating the total number of tulips in each rectangular field. Some students may count the number of tulips in one row, then add the same number for the second row, and so on (for example, 11, 22, 33, 44, and so on). Other students will use multiplication to find the number of tulips in 22 rows of 11 tulips each. Using these counting strategies will ensure that students will later have a better understanding of the area formula: $A = l \times w$, which is introduced in Section C.

Planning Students may work in pairs or in small groups on problem **3.** Encourage them to use more than one method to compare the areas. Be sure to discuss students' strategies.

Comments about the Problems

3. If some students count the tulips individually, suggest that they count the number of tulips in only one row and estimate or use multiplication to find the total number of tulips rather than count them individually.

Reasonable Prices

Mary Ann works at a craft store. One of her duties is to price different pieces of cork. She decides that 80¢ is a reasonable price for the big square piece (figure **a** below). Next she has to decide on the prices of the other pieces. Use **Student Activity Sheet 2** to solve the following problem.

4. What should the other prices be? *(Note:* All of the pieces have the same thickness.)

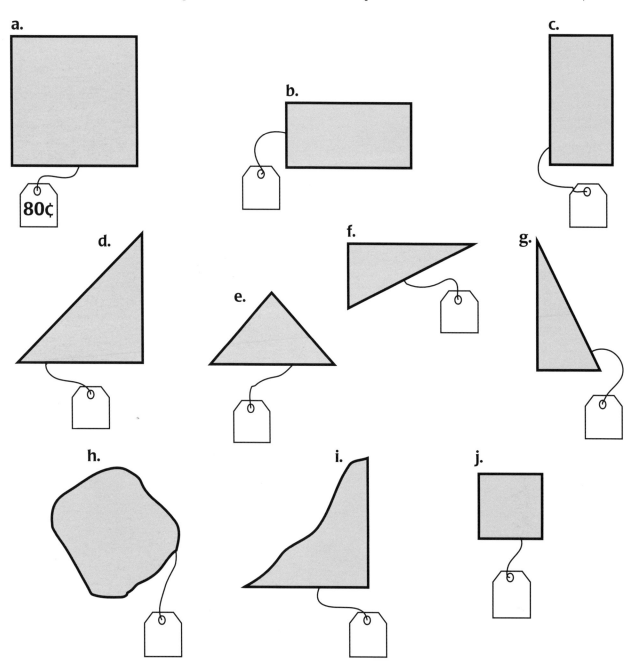

4. Estimates will vary. Sample student responses:

b. 40¢

c. 40¢

d. 40¢

e. 20¢

f. 20¢

g. 20¢

h. 40¢

i. 40¢

j. 20¢

Materials Student Activity Sheet 2 (one per student); tracing paper, optional (one sheet per student); scissors, optional (one pair per student)

Overview Students estimate the prices of pieces of cork of different shapes and sizes by comparing the area of each piece with the area of a square piece of cork priced at 80¢.

About the Mathematics In this activity, the concept of comparing areas of different shapes is extended to the concept of comparing prices of different-shaped items based on the area of each item. Students may estimate the areas of the triangular shapes without using a formula by comparing them to the areas of the rectangles and the square. The formula for finding the area of a triangle is made explicit in Section D.

Planning Tell students that they may use the square piece of cork labeled **a** as a reference point in order to estimate the prices of the other pieces. They may also use other pieces of cork as reference points. For example, cork **e** fits four times onto cork **a,** and two times onto cork **d.** Ask students to find as many relationships as they can. If time is a concern, at least discuss strategies for finding the prices of corks **e, h,** and **j.** Students may work in pairs or in small groups on problem **4.**

Comments about the Problems

4. If students are having difficulty, suggest that they trace the pieces of cork, cut them out, and overlay them onto piece **a** or onto any other piece of known price. Students should be able to reason about the areas of the different pieces using the fraction concepts they learned in the grade 5/6 unit *Some of the Parts.* For example, piece **j** fits four times onto piece **a,** so it is one-fourth of piece **a** and the price will be one-fourth of 80¢ (80¢ ÷ 4 = 20¢).

h. Students may use several strategies:
- compare the size of piece **h** to the size of piece **a** and give an estimate,
- piece **e** fits two times onto piece **h.**

Writing Opportunity Ask students to explain their suggested prices in their journals.

Dividing a Square

5. a. On a blank sheet of paper, draw six squares of the same size. Divide each square into eight equal parts. Use a different method for each square. The equal parts do not need to have the same shape.

b. Neysa drew this solution. Did she draw eight equal parts? Explain your answer.

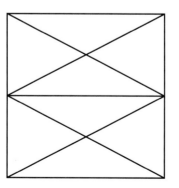

5. a. Sample student solutions:

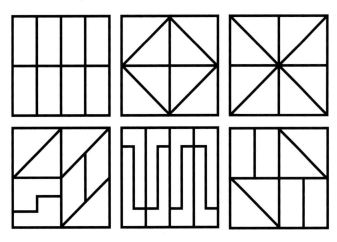

b. Yes. The parts are shaped differently, but each one covers the same amount of space. Sample student response:

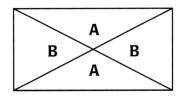

Two part A's cover the same space as two part B's.

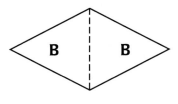

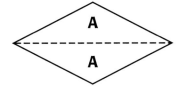

Materials drawing paper (one sheet per student); rulers (one per student); tracing paper, optional (one sheet per student); scissors, optional (one pair per student)

Overview Students use different methods to divide six squares into eight equal parts. The equal parts in each square may or may not have the same shape.

Planning Have students work in pairs or in small groups on problem **5** so they can discuss their strategies and solutions. Let students use a ruler to ensure that the squares are accurately divided into eight equal parts. When students have finished, you might display a variety of solutions on a bulletin board.

Comments about the Problems

5. a. Discuss the methods used to divide the squares into equal parts. Some students may have used paper-folding or tracing paper. Students may demonstrate that the parts are equal by tracing the shapes, cutting them out, and redistributing or reallotting them.

b. Some students may have produced this solution for problem **5a.** They may reason there are two different-shaped triangles in the square, so you need to show only that these two shapes cover the same area. Students may demonstrate that the pieces are equal by tracing the shapes, cutting them out, and fitting them together.

Below are drawings of tiles with different shapes. The small square tile costs $5. Use **Student Activity Sheet 3** to solve the following problem.

6. Figure out fair prices for the other tiles. Discuss your strategies with some of your classmates.

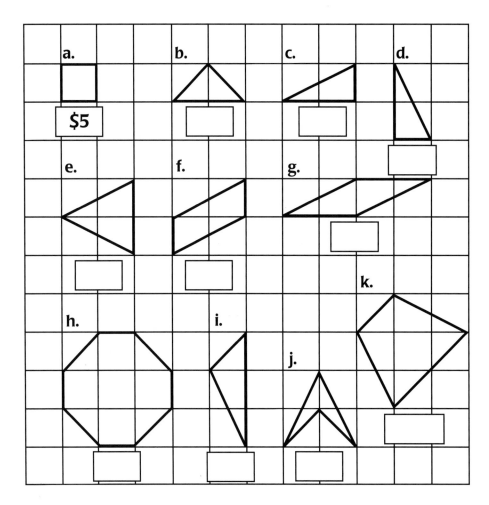

To figure out prices, you compared the sizes of the shapes to a square that costs $5. The square was the *measuring unit*. It is helpful to use a measuring unit when comparing sizes.

The number of measuring units needed to cover a shape is called the *area* of the shape.

6. The prices can be found by reshaping (transforming the shapes by cutting and pasting, and then comparing the shapes with that of the $5 tile).

b. $5

c. $5

d. $5

e. $10

f. $10

g. $10

h. $35

i. $7.50

j. $5

k. $22.50

Materials Student Activity Sheet 3 (one per student); scissors, optional (one pair per student); glue or tape, optional (one bottle or roll per pair or group of students)

Overview Students price tiles of different shapes and sizes by comparing their areas to the area of a $5 tile. This problem is similar to problem **4.**

About the Mathematics Students may use several strategies to find the area of a shape:

• reshape it by cutting and pasting, (see shape **b**);

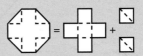

• divide shapes into other shapes, (this strategy can be used for shape **h**);

• take the sum or difference of two other shapes (for example the difference between shape **e** and shape **b** is **j**).

Using these strategies will help students later understand how to devise formulas to find the areas of different regular shapes.

Planning Students may work in pairs or in small groups on problem **6.** Be sure to discuss the different strategies used so students will begin to develop a repertoire of strategies. Problem **6** can be used as an assessment or assigned as homework.

Comments about the Problems

6. Informal Assessment This problem assesses students' ability to identify, describe, and classify geometric figures and their ability to compare the areas of shapes using a variety of strategies and measuring units. This problem may also be assigned as homework.

Most students will be able to visually reshape the figures or make drawings to compare the areas of the shapes. Other students may need to cut and paste.

Tessellations

When you tile a floor, wall, or counter, you want the tiles to fit together without a lot of space between them. Patterns without open spaces between the parts are called *tessellations*.

Sometimes you have to cut tiles to fit the edges or borders. The tiles in the pattern on the left fit together—they form a tessellation.

7. Using the $5 square (shown here again), estimate the price of each of the two tiles below.

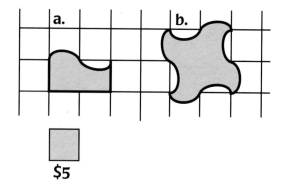

$5

Both of the tiles shown above can be used in tessellations. Use **Student Activity Sheet 4** to solve problem **8.**

8. Which of the tiles in problem **6** on page 5 can be used in tessellations? Draw enough of each pattern to show that it is a tessellation.

7. a. $10

  b. $20

8. All the shapes can be used in tessellations except for shape h (the octagon).

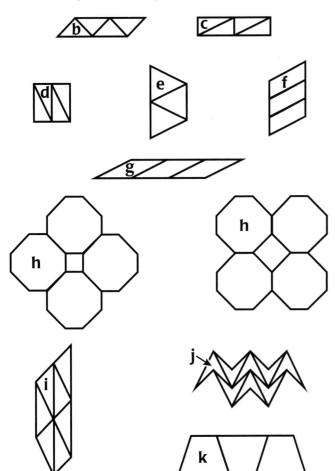

Materials Student Activity Sheet 4 (one per student); grid paper, optional (one sheet per student); tracing paper, optional (one sheet per student); transparency, optional (one per class); scissors, optional (one pair per student); compasses, optional (one per student); overhead projector, optional (one per class)

Overview Students estimate the prices of two tiles that can be used for a tessellation. They then investigate which of the tiles on Student Book page 5 can be used for a tessellation.

About the Mathematics Students are introduced to tessellations—congruent copies of a figure that do not overlap and have no open spaces. These patterns help students become aware of the concept of reallotment—a powerful tool for estimating area.

Tessellations develop students' understanding of the concepts of area and measuring units, which do not have to be squares. For example, using triangle **c** on Student Book page 5 as a measuring unit, the area of the rectangle below is four triangles.

Planning Students may work in pairs or in small groups on problem **8.** You may ask some students to draw their tessellations for problem **8** on a transparency and show their solutions to the class. This problem may also be assigned as homework.

Comments about the Problems

8. Homework This problem may be assigned as homework. The grid provided on Student Activity Sheet 4 has a different-sized square unit than that used on pages 5 and 6 of the Student Book. This will prevent students from simply copying or tracing the shape.

Students may have difficulty with the arrow (tile **j**) and the octagon (tile **h**). You may suggest that they draw about four of each of these shapes on a piece of grid paper and cut them out. Then students can investigate how the shapes will fit together.

Extension Ask students to draw an original tessellation. They should discover the only polygons that tessellate are rectangles, triangles, squares, and regular hexagons.

More about Tessellations

Tessellations often produce beautiful patterns. Artists from many cultures have used tessellations in their work. Here are some creations of the Dutch artist M.C. Escher.

9. Use grid paper or tracing paper to draw the basic design that is repeated in each of the patterns shown above.

9.

Materials tracing paper or grid paper (one sheet per student)

Overview Students look at the tessellations in two works of the Dutch artist M.C. Escher.

About the Mathematics Exploring tessellations emphasizes that measuring units for area may include shapes other than squares.

Planning Students may work in pairs or in small groups on problem **9.** Drawing the basic designs for tessellations may be time-consuming, so you may want to assign problem **9** as homework.

Comments about the Problems

9. **Homework** This problem may be assigned as homework. Some students may include a black-and-white pair of figures as the repeated design in each pattern. You may ask whether they would change their answer if they were to pay attention only to the shape of the repeating design and not its color.

Interdisciplinary Connection Have students find information on the works of the Dutch artist Maurits Cornelius Escher or on the tessellations found in tile patterns of early Greek or Roman architecture. Ask students to write brief reports about their findings. Some reports might be shared with the class.

Did You Know? M.C. Escher (1898–1972) was a Dutch graphic artist known for using realistic details to achieve strange visual effects in prints. He was a master of illusion—transforming and distorting images using mathematical tricks. His pictures were of equal interest to mathematicians, psychologists, and the general public and were widely reproduced in the mid-20th century.

Here is one way to make tessellations. Start with a rectangular tile and change the shape according to the following rule:

What is changed in one place must be made up for elsewhere.

For instance, if you add a shape onto the tile like this,

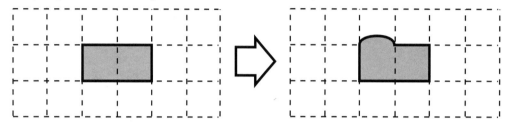

you have to take away the same shape someplace else. Here are a few possibilities.

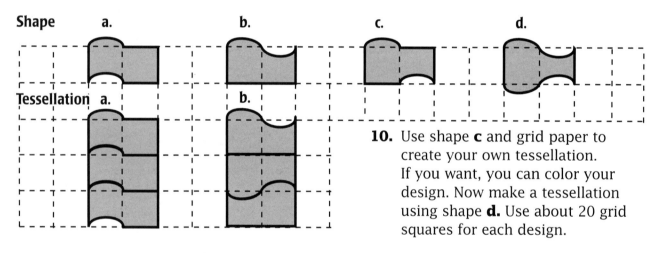

Shape   **a.**   **b.**   **c.**   **d.**

Tessellation **a.** **b.**

10. Use shape **c** and grid paper to create your own tessellation. If you want, you can color your design. Now make a tessellation using shape **d.** Use about 20 grid squares for each design.

Shape **d** looks like a fish. You can add a mouth, but then you must take away part of the tail.

Here are some fish with eyes and some with contrasting colors.

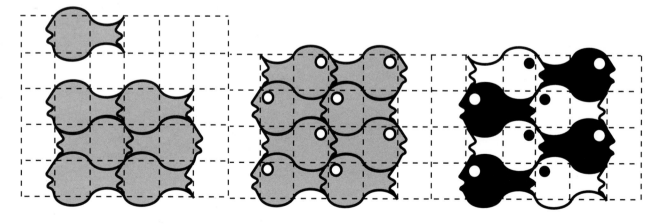

10. Tessellations should look like those below.

Shape

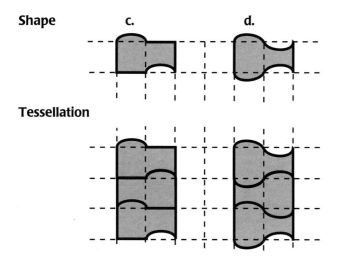

c.   d.

Tessellation

Materials grid paper (one sheet per student); colored pencils, optional (one box per student)

Overview Students make new tessellation designs by cutting, pasting, and rearranging parts of existing rectangular tiles.

About the Mathematics When a rectangular tile used as a basic tessellating unit is altered by subtracting a portion from one place and adding it to another, the area remains the same.

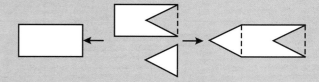

Planning Students may work in pairs or in small groups on problem **10.** Creating tessellations can be time-consuming, so you may want to assign problem **10** as homework. You may have students use colored pencils to decorate their designs. Posters of Escher drawings, as well as students' designs, might make an attractive display in the classroom.

Comments about the Problems

10. Homework This problem may be assigned as homework. Make sure students understand that even after they change the shape of the original rectangular tile, the tile's area remains the same. To demonstrate this concept, ask students to compare the areas of the four tiles **a** through **d.** They should be able to see and explain that the areas of the four tiles are equal.

Extension The pictures on this page also reinforce the idea that area can be calculated using measuring units other than squares. Ask students to find the area of tessellation **a** using both a grid square as a measuring unit (6 squares) and using tile **a** as a measuring unit (3 tiles). Then ask students to determine the area of the fish-shaped tessellation pattern without mentioning the measuring unit. Some students may say that they need to know the measuring unit. Some students may say the area can be different depending on the measuring unit you use. (The area is 8 units using a fish tile as the measuring unit; the area is 16 units using one square as a measuring unit.)

The trick in making tessellations is to create the pattern little by little.

Step 1

Step 2

Step 3

Step 4

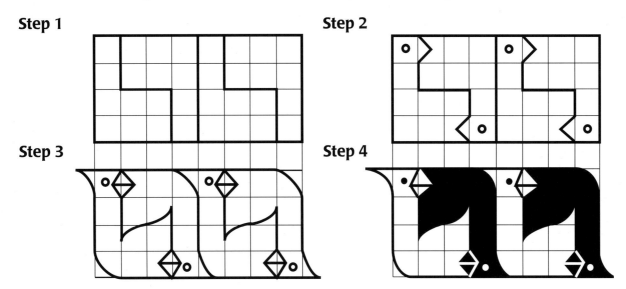

Here are other examples.

Example A

Example B

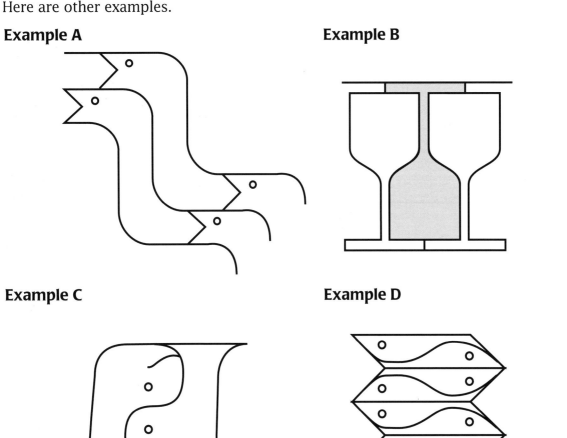

Example C

Example D

11. Make your own tessellations on grid paper. Draw enough of each pattern to show that it is a tessellation.

11. Tessellations will vary. Sample tessellations:

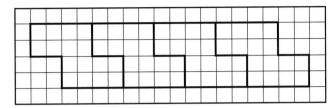

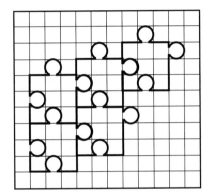

Materials grid paper (one sheet per student); scissors, optional (one pair per student); glue or tape, optional (one bottle or roll per pair or group of students)

Overview Students see other examples of tessellations. They then create their own tessellation.

Planning Students may work in pairs or in small groups on problem **11.** Problem **11** may also be used as an assessment and/or assigned as homework. Make sure that students understand the steps involved in making tessellations before they begin their designs. You might display students' original tessellations in the classroom.

Comments about the Problems

11. Informal Assessment This problem assesses students' ability to create and work with tessellation patterns. It may also be assigned as homework.

If students are having difficulty getting started, suggest that they begin with a simple shape, such as a rectangle. They can then cut and paste parts of the rectangle according to the steps described on Student Book page 8. When students finish their basic design, they can draw their tessellation on a new piece of grid paper.

Extension Students could present their tessellations to the class and ask their classmates to identify the basic shape used and to explain how that basic shape was created. The designer of the tessellation can also challenge classmates to find the area of the basic tessellation unit in his or her design.

Summary

This section is about comparing areas (sizes) of shapes. You used many different methods to compare shapes. You counted dots and tulips. You estimated how many times different-shaped pieces of cork would fit onto a given square piece. You divided shapes and put shapes together to make new shapes. You also looked at and made your own tessellation patterns (patterns made with pieces that fit together with no open spaces between them).

Summary Questions

12. A 9 inch × 13 inch rectangular cake has been cut into three pieces, as shown in the picture on the far right. If the whole cake costs $3.60, how much should each piece cost?

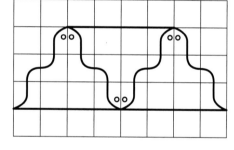

Alex made the two tessellation patterns on the right. First he drew a shape on grid paper that used whole squares. Then he changed it in one place and made up for the change in another place.

Ghost Tessellation

13. Answer these questions about one of Alex's tessellations.

   a. How many squares did Alex start with to make the basic shape? How did he make the shape?

   b. How did Alex make the tessellation?

   c. How many squares make up one ghost or one arrow? (Another way to ask this question is "What is the area of one ghost or one arrow measured in squares?")

Arrow Tessellation

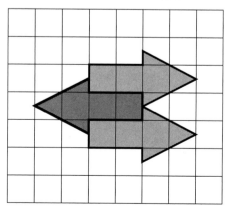

12. The rectangular piece of cake will cost $2.40, and the two triangular pieces of cake will cost 60 cents each.

Sample student response:

The two triangles at the bottom form a 3-by-13-inch piece. I divided the 6-by-13-inch piece in half. Now I have three 3-by-13-inch pieces. Since the whole cake costs $3.60, each 3-by-13-inch piece will cost $1.20. A triangular piece is half of a 3-by-13-inch piece, so it costs 60¢.

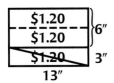 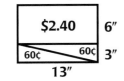

13. a. ghost: 6 squares
arrow: 4 squares

   b. Ghost: Alex made the ghost by starting with a rectangle that was 2 squares by 3 squares. He drew the ghost's head in the middle of the top four squares. He took the leftovers and rotated them to make the shoulders, as shown below.

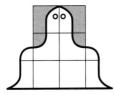

To make the tessellation, Alex put the first ghost right side up and the next ghost upside down.

Arrow: Alex made the arrow by starting with a rectangle that was four squares long. He drew a point at one end. He took the leftover triangles and turned them over. Then he added these leftovers to the sides of the next square, as shown below.

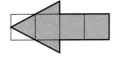

To make the tessellation, Alex put the arrows in rows, facing opposite directions.

   c. ghost: 6 squares
arrow: 4 squares

Materials tracing paper, optional (one sheet per student); grid paper, optional (one sheet per student)

Overview Students read the Summary, which reviews the different methods used in this section to determine the areas of different shapes. Students then find the prices for different-sized pieces of cake, given the area and cost of the entire cake. They also answer questions about two new tessellations.

Planning Students may work individually or in pairs on problems **12** and **13**. Problem **12** may be assigned as homework. Problem **13** may be used as an assessment. After students complete Section A, you may assign appropriate activities from the Try This! section, located on pages 49–52 of the Student Book, as homework.

Comments about the Problems

12. Homework This problem may be assigned as homework. Students may use different strategies to find the prices of the cake pieces relative to that of the whole cake. Some students may divide the rectangular shape into six equal triangles and then divide the total price by six to determine the cost of each piece. (See the solutions column.)

13. Informal Assessment This problem assesses students' ability to create and work with tessellation patterns.

   a. If students have trouble explaining where the basic shape came from, tell them that Alex started with a 2 × 3 rectangle for the ghost and a 4 × 1 rectangle for the arrow.

   c. Informal Assessment This problem assesses students' ability to estimate and compute the areas of geometric figures.

Work Students Do

In this section, students further develop their understanding of the concept of area. They first compare the areas of the states of Texas, Utah, and California. Students then use a rectangular grid to compare and estimate the areas of two imaginary islands. They solve a problem about trading pieces of land using the strategy of reallotment. Students compare and estimate areas of different geometric shapes using a variety of strategies. They also determine the prices of different-sized pieces of felt using the given dimensions and price of one piece. Finally, students find the areas of triangles enclosed in rectangles.

Goals

Students will:

- identify, describe, and classify geometric figures;*

- compare the areas of shapes using a variety of strategies and measuring units;

- estimate and compute the areas of geometric figures;

- understand which units and tools are appropriate to estimate and measure area, perimeter, and volume;*

- use the concepts of perimeter, area, and volume to solve realistic problems;

- represent and solve problems using geometric models.

  * This goal is assessed in other sections in the unit.

Pacing

- approximately four or five 45-minute class sessions

Vocabulary

- contiguous

About the Mathematics

When a rectangular grid is used to compare and estimate the areas of shapes, one square in the grid is used as a measuring unit. To find the number of squares that cover a shape, a variety of strategies can be used: counting the number of whole squares in a shape and estimating the number of squares that the remaining pieces will make; subdividing the shape into parts that are easy to estimate; reshaping the figure by cutting and pasting so that the area of the new shape can be found easily (shown below left); and enclosing the shape in a rectangle and subtracting the areas outside the shape (shown below right).

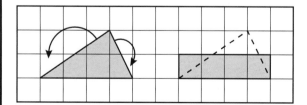
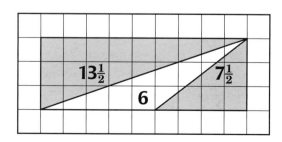

Materials

- Student Activity Sheets 5–8, pages 129–132 of the Teacher Guide (one of each per student)
- almanacs, page 29 of the Teacher Guide, optional (one per pair or group of students)
- tracing paper, page 29 of the Teacher Guide, optional (one sheet per student)
- grid paper, page 29 of the Teacher Guide, optional (one sheet per student)
- scissors, page 29 of the Teacher Guide, optional (one pair per student)
- transparency of Student Activity Sheet 6, page 31 of the Teacher Guide, optional (one per class)
- transparency of maps on Student Book page 12, page 31 of the Teacher Guide, optional (one per class)
- transparency of Student Activity Sheet 7, page 33 of the Teacher Guide, optional (one per class)
- transparency of Student Book page 15, page 37 of the Teacher Guide, optional (one per class)
- overhead projector, pages 31, 33, and 37 of the Teacher Guide, optional (one per class)
- Sizing up Islands assessment, page 137 of the Teacher Guide, optional (one per student)

Planning Instruction

It is important for students to go beyond the method of merely counting squares to find the area of a shape. Encourage students to use different methods to determine area and to share their strategies with each other. These strategies for calculating area will be made more explicit in Section C. In particular, discuss students' strategies for solving problems 1b, 2, 5, 7, 8, 10, 11, 12, and 14. The Sizing up Islands assessment can be used after students complete Section B.

Students can work on problem 11 individually or in pairs. Students can work on the remaining problems in this section in pairs or in small groups.

There are no optional problems in this section.

Homework

Problem 12 (page 36 of the Teacher Guide) and the Extensions (pages 29 and 35 of the Teacher Guide) can be assigned as homework. After students complete Section B, you may assign appropriate activities from the Try This! section, located on pages 49–52 of the Student Book. The Try This! activities reinforce the key mathematical concepts introduced in this section.

Planning Assessment

- Problem 8 can be used to informally assess students' ability to represent and solve problems using geometric models and their ability to estimate and compute the areas of geometric figures.
- Problem 9 can be used to informally assess students' ability to compare the areas of shapes using a variety of strategies and measuring units.
- Problem 10 can be used to informally assess students' ability to estimate and compute the areas of geometric figures.
- Problem 11 can be used to informally assess students' ability to estimate and compute the areas of geometric figures and to use the concepts of perimeter, area, and volume to solve realistic problems.
- The Sizing up Islands assessment assesses students' ability to use the concepts of perimeter, area, and volume to solve realistic problems; to understand which units and tools are appropriate to estimate and measure area, perimeter, and volume; and to compare the areas of shapes using a variety of strategies and measuring units.

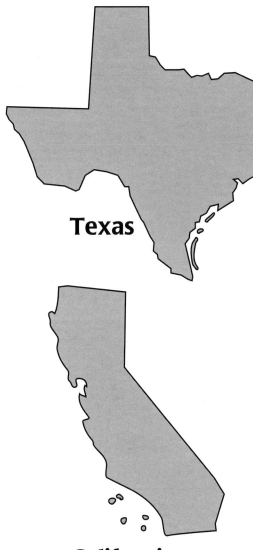

Texas

California

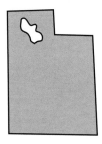

Utah

Big States, Small States

Designers of tourism advertisements often expect people to recognize the shape of a state.

1. **a.** Without looking at a map, draw the shape of the state in which you live.

    b. If you were to list the 50 states from the largest to the smallest in land size, where would you rank your state?

On this page, three U.S. states are drawn to the same scale.

2. Estimate to answer the following questions and explain how you found each estimate.

    a. How many Utahs fit into California?

    b. How many Utahs fit into Texas?

    c. How many Californias fit into Texas?

3. Imagine a map of the United States drawn to the same scale as the states on this page. How would you change the drawing you made of your state to fit on this imaginary map?

Forty-eight of the United States are *contiguous*, or physically connected. You will find the drawing of the contiguous states on **Student Activity Sheet 5.**

4. Compare the area of your state to the area of the 48 contiguous United States.

1. a. Drawings will vary.

   b. Answers will vary.

2. a. About two Utahs would fit into California. Accept a range from $1\frac{1}{2}$ to $2\frac{1}{2}$. Explanations will vary. Some students may actually trace and cut out the area of Utah and lay it on California as shown below.

   b. About three Utahs would fit into Texas. Accept a range from $2\frac{1}{2}$ to $3\frac{1}{2}$. Explanations will vary. Some students may trace and cut, as shown below.

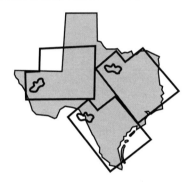

   c. Almost two Californias would fit into Texas. Accept a range from $1\frac{1}{2}$ to $2\frac{1}{2}$. Explanations will vary. Some students may trace and cut as shown below.

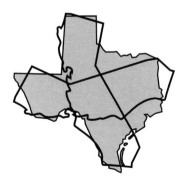

3. Answers will vary depending on the original size of the state that students drew in problem **1a.** An approximation is required here.

4. Answers will vary depending on the estimation strategies students use.

Materials Student Activity Sheet 5 (one per student); almanacs, optional (one per pair or group of students); tracing paper, optional (one sheet per student); grid paper, optional (one sheet per student); scissors, optional (one pair per student)

Overview Students compare the areas of three states and then draw the shape of their own state using the same scale. Students use a U.S. map to compare the area of their state to the total area of the 48 contiguous states.

About the Mathematics Problems **2–4** are powerful illustrations of what is meant by the term *area*: the number of measuring units needed to cover an entire region.

Planning Students may work on problems **1–4** in pairs or in small groups. After students finish problem **2,** discuss their solutions and strategies for problems **1b** and **2.** You may want to orient students to the U.S. map. Be sure they understand the expression *contiguous states.* Discuss students' strategies for problem **4.**

Comments about the Problems

2. Encourage students to devise their own strategies. For parts **a** and **b,** some students may try to cover the other states with as many Utahs as possible. Others may use grid paper to estimate the area of each state and then make relative comparisons.

   c. Some students may estimate the areas of both states while others may use their answers to parts **a** and **b.**

4. Avoid standard units at this stage unless initiated by students.

Extension Have students use information from almanacs or other resource material to compare the area ranking of the state in which they live to its population ranking. Ask students the following question before they see the ranking lists: *Do you think the order of states ranked according to size will be the same order of states when ranked by population?* [No] *Why?* [Population ranking has no relation to the area ranking.]

Islands

Here are two islands: Smoke Island and Fish Island. Use **Student Activity Sheet 6** to solve the problems below.

Smoke Island

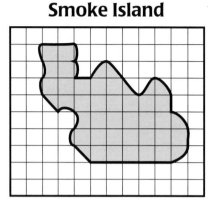

Fish Island

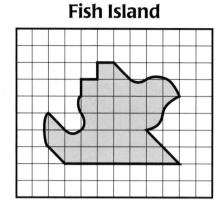

5. **a.** Which island is bigger? How do you know?

    b. Estimate the area of each island.

Theme Park

Fantasy World would like to build a theme park next to a wildlife preserve. To do that they must persuade two farmers to trade a piece of land next to the wildlife preserve for one that the company owns that used to be the site of a factory.

Below are two maps of the situation.

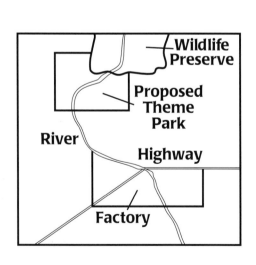

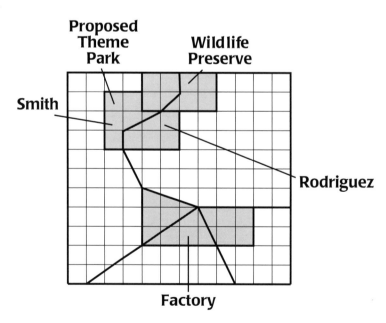

6. **a.** Describe the differences between the two maps.

    b. Do you think the map on the right provides reasonable estimates of the actual areas? Explain.

5. a. Smoke Island is bigger than Fish Island. Explanations will depend on the strategies students used. Sample student response:

For each island, I counted the number of complete squares covered. Then I combined the leftover parts of squares to make complete squares and counted them.

b. Estimates will vary. For Smoke Island, accept estimates in the range of 35 to 41 square units. For Fish Island, accept estimates in the range of 28 to 34 square units.

6. a. Answers will vary. Sample student response:

The map on the left shows the curves in the river and in the boundary of the wildlife preserve. The map on the right has straight lines and a grid.

b. Answers will vary. Sample student response:

The map on the right gives reasonable estimates of the actual areas. Except for the wildlife preserve and the river, the boundaries were already straight. The wildlife preserve could be changed into a rectangle by cutting and pasting.

Materials Student Activity Sheet 6 (one per student); transparency of Student Activity Sheet 6, optional (one per class); overhead projector, optional (one per class); transparency of the maps on Student Book page 12, optional (one per class)

Overview Students compare the areas of two islands and use different strategies to decide which island is the largest. They then are introduced to the concept of the reallotment of land. Students compare two maps representing the same area and describe the differences between them.

Planning Discuss the maps on Student Book page 12 to ensure that students understand the information shown. Ask them to describe what they see on each map. Students may work in small groups on problems **5** and **6** so that they can share their ideas. After they finish, briefly discuss their solutions.

Comments about the Problems

5. Encourage students to estimate the areas of the islands using the grid squares. In Section A, students reallotted shapes by subtracting one part and adding to another. Here they are comparing the areas of two different shapes. If students are having difficulty estimating the areas, make a transparency of Student Activity Sheet 6 to show students' different counting strategies. Students may devise their own strategy or use one of the strategies mentioned in the About the Mathematics section on page 26 of the Teacher Guide.

6. You may want to have students show their explanations on an overhead projector using a transparency of the two maps.

The proposed location for the theme park is divided into two sections by a river. One piece of land is owned by a farmer named Smith, and the other by a farmer named Rodriguez.

7. What is the area of each farmer's land? Use the scale in the picture below to help you estimate.

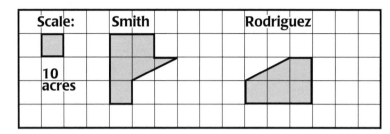

Smith: _?_ squares = _?_ acres   Rodriguez: _?_ squares = _?_ acres

The factory site is divided into three pieces by a highway and a river.

8. Estimate the area of each piece in squares and in acres.

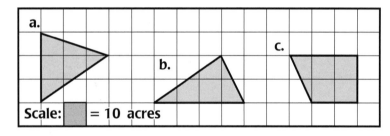

9. Do you think the farmers will be willing to trade their land for the factory site? Why or why not?

Use **Student Activity Sheet 7** to answer problem **10.**

10. Determine the area of each of the shaded pieces on the right. Give your answers in square units. Be prepared to explain your reasoning.

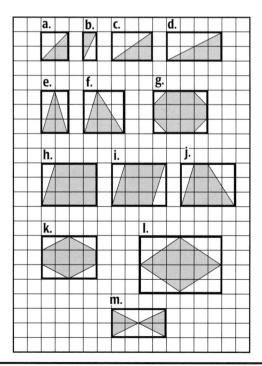

7. Area of Smith's land, 5 squares = 50 acres
Area of Rodriguez's land, 5 squares = 50 acres

8. a. 4.5 squares = 45 acres

b. 4 squares = 40 acres

c. 5 squares = 50 acres

9. Answers will vary. Some students may reason that the total area of the factory site is greater than that of the farmers' land, so the farmers would trade. Other students may compare the features of the farmers' land with those of the factory site.

Sample student responses:

- The farmers may not trade their land because the factory land has poor soil.

- The farmers like being near a wildlife preserve.

- The farmers might be willing to trade their land of 50 acres each if they know how the factory land of 145 acres will be divided.

10. a. 2 square units

b. 1 square unit

c. 3 square units

d. 4 square units

e. 3 square units

f. $4\frac{1}{2}$ square units

g. 10 square units

h. $10\frac{1}{2}$ square units

i. 9 square units

j. $7\frac{1}{2}$ square units

k. 8 square units

l. 12 square units

m. 4 square units

Explanations will depend on the strategies students used. Students' reasoning should show the methods used to reallot each figure to find its area.

Materials Student Activity Sheet 7 (one per student); transparency of Student Activity Sheet 7, optional (one per class); overhead projector, optional (one per class)

Overview Students determine the areas of different pieces of land. They develop arguments for or against the trading of certain parcels of land. Students also estimate the areas of different geometric figures.

About the Mathematics To find the area of the triangles on Student Activity Sheet 7, some students may use the rectangular frame formed by the grid lines that enclose each triangle. They can count the number of squares that make up the area of the rectangle and halve that amount to determine the area of the triangle.

Planning Students may work in pairs or in small groups on problems **8–10.** Discuss students' answers and strategies for one of the shapes in problem **7** and shape **a** in problem **8.** Problems **8–10** may be used as assessments.

Comments about the Problems

8. Informal Assessment This problem assesses students' ability to estimate and compute the areas of geometric figures. It also assesses their ability to represent and solve problems using geometric models. Students may use a variety of strategies.

9. Informal Assessment This problem assesses students' ability to compare the areas of shapes using a variety of strategies and measuring units. Strategies may vary.

10. Informal Assessment This problem assesses students' ability to estimate and compute the areas of geometric figures.

Finding the areas of figures **a–d** can involve the strategy of halving the area of a rectangle. Some students may devise and use a subtraction strategy for figures **g, i, k, l,** and **m.** Make sure that students include units (squares or square units are acceptable).

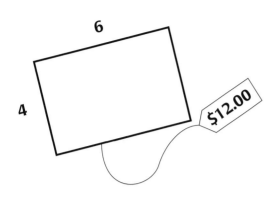

6

4

$12.00

The students in Ms. Petry's class plan to purchase felt in different colors to make a wall hanging. The felt comes in sheets 4 feet by 6 feet. Each sheet costs $12. The salespeople have agreed to cut the felt for the students and to charge them for only the amount of felt they need. The shaded sections shown below are the pieces the students would like to purchase. Use **Student Activity Sheet 8** to solve problem **11.**

11. Calculate the prices of the shaded pieces.

a.
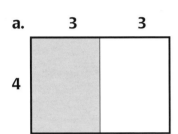

b.
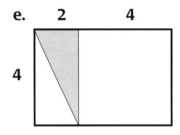

c.
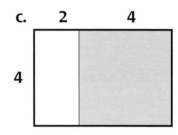

d.
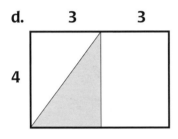

e.
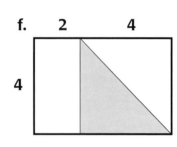

f.
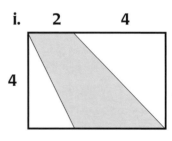

g.
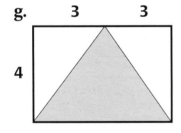

h.
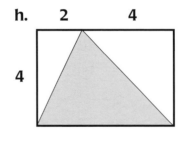

i.

j.
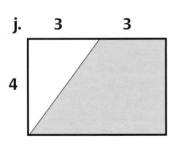

k.
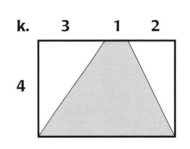

l.
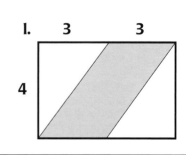

11. a. $6 ($\frac{1}{2}$ of $12)

   b. $4 ($\frac{1}{3}$ of $12)

   c. $8 ($\frac{2}{3}$ of $12). *Note:* The diagram printed in the Student Book contains a shaded area that is incorrect. The one in this Teacher Guide is correct, as is the one on Student Activity Sheet 8.

   d. $3 ($\frac{1}{2}$ of the area of figure **a**)

   e. $2 ($\frac{1}{2}$ of the area of figure **b**)

   f. $4 ($\frac{1}{2}$ of the area of figure **c**). *Note:* The diagram printed in the Student Book contains a shaded area that is incorrect. The one in this Teacher Guide is correct, as is the one on Student Activity Sheet 8.

   g. $6 (twice the area of figure **d**)

   h. $6 (the area of figure **e** plus the area of figure **f**)

   i. $6 (the same as the area of figure **h**)

   j. $9 (the area of figure **a** plus the area of figure **d**)

   k. $7 (the area of figure **d** plus the area of figure **e** plus $\frac{1}{2}$ the area of figure of **b**)

   l. $6 (twice the area of figure **d**)

Materials Student Activity Sheet 8 (one per student)

Overview Students determine the prices of different-sized pieces of felt using the dimensions and price of one piece.

About the Mathematics To find the area of the shaded pieces, students may use one of the following strategies: halving; subtraction; constructing a grid; relating one problem to another; or dividing each diagram into a series of smaller rectangles and triangles, calculating the areas of these smaller shapes, and sometimes adding areas to equal the shaded area.

Planning Students can work individually or in pairs on problem **11**. This problem can be used as an assessment.

Comments about the Problems

11. Informal Assessment This problem assesses students' ability to estimate and compute the areas of geometric figures. It also assesses students' ability to use the concepts of perimeter, area, and volume to solve realistic problems.

Here is a sample strategy for figure **k:**
- Divide the large rectangle into three smaller rectangles (one 3 × 4, one 1 × 4, and one 2 × 4), as shown below.

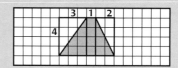

- Calculate the cost of the shaded area of each rectangle. The center rectangle is completely shaded, so its area is 1 × 4 = 4 square units. Its cost is $2, half the cost of figure **b.** The shaded area of the rectangle on the left is half of its total area (6 square units). Its cost is $3, the same as figure **d.** The shaded area of the rectangle on the right is also half of its total area (4 square units). Its cost is $2, the same as figure **e.**

Extension You might ask students if they notice any relationship between the figures in the first and second rows. Figures **d, e,** and **f** are half the size of figures **a, b,** and **c** respectively. This may help students to develop a strategy for finding the area of a triangle.

12. What is the area (in square units) of each of the shaded pieces below?

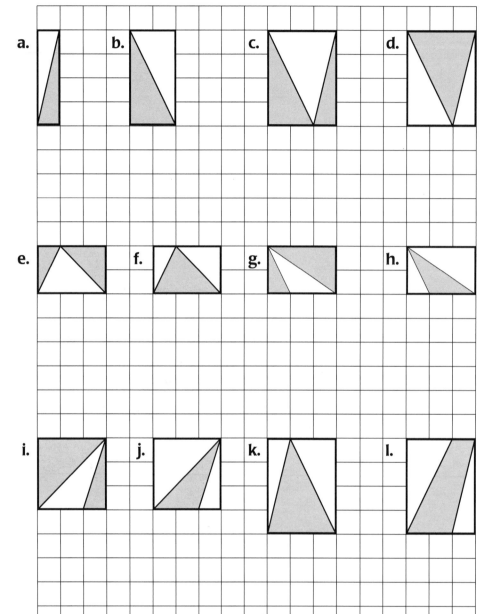

12. a. 2 square units

   b. 4 square units

   c. 6 square units

   d. 6 square units

   e. 3 square units

   f. 3 square units

   g. 4 square units

   h. 2 square units

   i. 6 square units

   j. 3 square units

   k. 6 square units

   l. 6 square units

   m. 3 square units

   n. 3 square units

   o. 3 square units

   p. 3 square units

Note: Some students might use the subtraction strategy to find the areas of figures **o** and **p,** as shown below.

o. Area of the rectangle = 3 × 3 = 9 square units
Area of unwanted sections = $4\frac{1}{2} + 1\frac{1}{2}$ = 6 square units
Area of triangle = 9 − 6 = 3 square units

p. Area of the rectangle = 4 × 3 = 12 square units
Area of unwanted sections = 6 + 3 = 9 square units
Area of triangle = 12 − 9 = 3 square units

Materials transparency of page 15 of the Student Book, optional (one per class); overhead projector, optional (one per class)

Overview Students calculate the areas of triangles.

About the Mathematics This activity leads students closer to discovering a method or formula for the area of a triangle. The formula itself should not be introduced at this point. It is important, however, that students see that the area of a right triangle is exactly one-half of the area of a rectangle that has sides of the same lengths, such as in figures **a** and **b.** This concept can also be used to find the area of other triangles, as show below.

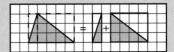

By subtracting the unshaded areas from the total area of the rectangle, the area of the shaded triangle is found. This strategy is especially useful for finding the areas of triangles like figure **h.** Note that the statement *the area of a triangle is one-half of the area of a rectangle* is not always valid. The triangle and rectangle in figure **p** illustrate this.

Planning Students may work on problem **12** in pairs or in small groups. Problem **12** may be assigned as homework. During the next class session, have students share their strategies with the class.

Comments about the Problems

12. Homework This problem may be assigned as homework. If students simply count the squares in the triangles, encourage them to use a strategy such as halving or subtracting unshaded portions. Some students may see that the area of figure **c** can be used to find the area of figure **d,** with the subtraction strategy.

Students may see and use other relationships between the shaded and unshaded areas of different figures. For example, each triangle in the last row has a base of 2 units, a height of 3 units, and an area of 3 units.

Summary

In this section, you compared the areas of various states, islands, pieces of land, and pieces of felt. You also measured and described areas using square units.

You explored several strategies for determining the areas of various shapes: for example, counting the number of complete squares inside a shape, then reallotting the remaining pieces to make new squares.

Inside this shape there are four complete squares.

The pieces that remain can be combined into four new squares as shown below.

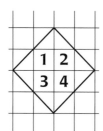

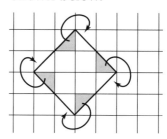

You may also have enclosed a shape with a rectangle and subtracted the empty areas.

Summary Questions

13. How else might you calculate the area of the shape in the Summary box above?

Here are four triangles.

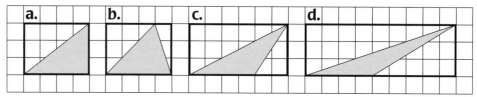

14. What similarities can you find in these four triangles?

15. What are the areas of two of the triangles? Explain how you found the areas.

13. Answers will vary.

Students may see the shape as four triangles, which can be reshaped to make two squares, each with an area of four square units, as shown below.

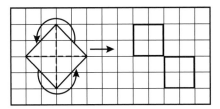

Other students may enclose the square in a larger square and recognize that the area of the smaller square is half the area of the larger square, as shown below.

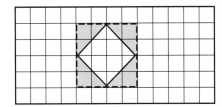

14. a. Answers will vary. Sample student responses:

- All the triangles have a height of 3 units.

- All the triangles have a base of 4 units.

- All the triangles have an area of 6 square units.

15. The area of each triangle is 6 square units. Explanations will vary. To find the area of triangle **d,** some students may use a subtraction strategy, as shown below.

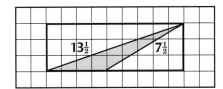

Area of rectangle: $3 \times 9 = 27$ square units.
Area of unwanted sections:
$7\frac{1}{2} + 13\frac{1}{2} = 21$ square units
Area of triangle $= 27 - 21 = 6$ square units

Other students may reallot portions of the triangle. Some students may reason that since the base is 4 units and the height is 3 units, the area will be half of 12, or 6 square units.

Materials Sizing up Islands assessment, optional (one per student)

Overview Students read the Summary, which reviews some of the strategies used to compare and determine the areas of various shapes, such as counting the complete squares in a figure and reallotting the remaining pieces to find the total number of squares. Students share additional strategies for determining the area of a square. They also look for similarities in four triangles and find the area of each.

Planning Students may work in pairs or in small groups on problem **14.** Discuss students' solutions and strategies for problem **14** with the whole class. After students complete Section B, you may assign appropriate activities from the Try This! section, located on pages 49–52 of the Student Book, as homework.

Comments about the Problems

14. Students will probably not use the words *base* and *height*. Accept any words that convey these ideas. These terms are introduced in Section C. Some students may reason that since the base of each triangle is 4 units and the height of each triangle is 3 units, the area will be half of 12 units, or 6 square units.

Assessment Opportunity The Sizing up Islands assessment can be given after students finish Section B. It gives them an opportunity to connect area to actual places on Earth. This activity assesses students' ability to use the concept of area to solve realistic problems; to understand which units and tools are appropriate to estimate and measure area; and to compare areas of shapes using a variety of strategies and measuring units.

Work Students Do

Students transform a parallelogram into a rectangle in order to find its area. They use different strategies to reshape a parallelogram, such as cutting and pasting one triangular section of the parallelogram to make a rectangle.

Students determine the areas of other quadrilaterals, such as rectangles and trapezoids, and investigate whether there is a rule for finding the area of a quadrilateral whose corners touch the sides of a rectangle. They focus on base and height when they transform a rectangle into various parallelograms. Students create different shapes having equal areas. They generalize rules or formulas for finding the areas of rectangles, parallelograms, and triangles using the words *base* and *height*. Finally, students use measuring units of different sizes to investigate the areas of shapes.

Goals

Students will:

- compare the areas of shapes using a variety of strategies and measuring units;
- create and work with tessellation patterns;
- estimate and compute the areas of geometric figures;*
- understand which units and tools are appropriate to estimate and measure area, perimeter, and volume;
- use the concepts of perimeter, area, and volume to solve realistic problems;
- represent and solve problems using geometric models;
- generalize formulas and procedures for determining the areas of rectangles, triangles, parallelograms, quadrilaterals, and circles.

* This goal is assessed in other sections in the unit.

Pacing

- approximately four 45-minute class sessions

Vocabulary

- base
- height
- parallelogram
- quadrilateral

About the Mathematics

Area is a measure of the space that a two-dimensional figure encloses. The traditional unit of measure is a square; however, space can also be measured using triangles, hexagons, or other shapes that can be tessellated.

The strategies for estimating and calculating area that students developed in Section B are made more explicit in this section. The areas of rectangles, parallelograms, and triangles can be found with counting, reallotting, and subtracting strategies. The use of base and height measurements leads to formulas for the areas of parallelograms and triangles. To derive a formula for the area of a parallelogram from the area of a rectangle, several strategies can be used, such as reshaping by cutting and pasting or shifting. The area of a triangle can be found with a subtraction strategy or by halving the area of a corresponding parallelogram, as illustrated below.

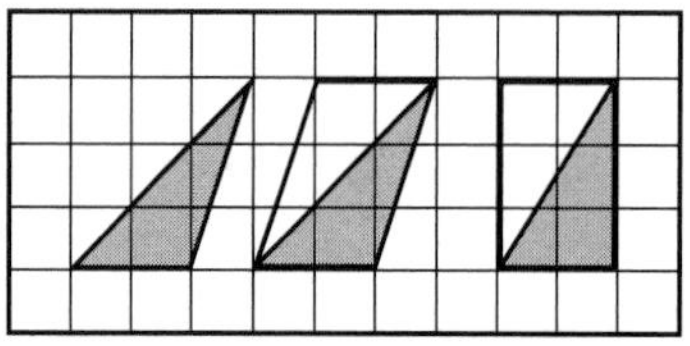

Materials

- Student Activity Sheets 9–11, pages 133–135 of the Teacher Guide (one of each per student)
- transparency of page 17 of the Student Book, page 43 of the Teacher Guide, optional (one per class)
- overhead projector, pages 43, 49, 51, 53, 57, and 63 of the Teacher Guide, optional (one per class)
- grid paper, pages 45, 49, 51, 53, and 65 of the Teacher Guide (one or two sheets per student)
- rulers, pages 47 and 51 of the Teacher Guide, optional (one per student)
- scissors, pages 49, 51, 53, 57, and 65 of the Teacher Guide (one pair per student)
- glue or tape, pages 49, 51, 53, 55, 57, and 65 of the Teacher Guide (one bottle or roll per pair or group of students)
- transparency of grid paper, pages 49, 51, 53, and 63 of the Teacher Guide, optional (one per class)
- transparency of Student Activity Sheet 9, page 57 of the Teacher Guide, optional (one per group of students)
- colored pencils, page 59 of the Teacher Guide, optional (one box per group of students)
- transparency of Figure 1 with an overlaid grid, page 63 of the Teacher Guide, optional (one per class)

Planning Instruction

Allow students to discuss their ideas for finding areas in order to develop their ability to understand and use other strategies.

You may want students to work individually on problems 4, 5–9, 13, 14, 16, 18, and 25. They may work in pairs or in small groups on the remaining problems in this section.

Problems 9 and 23 are optional. If time is a concern, you may omit these problems or assign them as homework.

Homework

Problems 3 (page 44 of the Teacher Guide), 14 (page 52 of the Teacher Guide), 15 (page 54 of the Teacher Guide), 17 (page 56 of the Teacher Guide) and 19–23 (page 58 of the Teacher Guide) can be assigned as homework. The Extensions (pages 59 and 61 of the Teacher Guide) may also be assigned as homework. After students complete Section C, you may assign appropriate activities from the Try This! section, located on pages 49–52 of the Student Book. The Try This! activities reinforce the key mathematical concepts introduced in this section.

Planning Assessment

- Problems 4, 5, 6, and 13 can be used to informally assess students' ability to generalize formulas and procedures for determining the areas of rectangles, triangles, parallelograms, quadrilaterals, and circles.
- Problem 9 can be used to informally assess students' ability to compare the areas of shapes using a variety of strategies and measuring units.
- Problems 14 and 25 can be used to informally assess students' ability to estimate and compute the areas of geometric figures and to understand which units and tools are appropriate to estimate and measure area, perimeter, and volume. Problem 14 also assesses students' ability to use the concepts of perimeter, area, and volume to solve realistic problems; and to compare the areas of shapes using a variety of strategies and measuring units.
- Problem 16 can be used to informally assess students' ability to represent and solve problems using geometric models.
- Problem 18 can be used to informally assess students' ability to create and work with tessellation patterns; to use the concepts of perimeter, area, and volume to solve realistic problems; and to understand which units and tools are appropriate to estimate and measure area, perimeter, and volume.

Quadrilateral Patterns

A *quadrilateral* is a four-sided figure. Each of the figures below is a special type of quadrilateral called a parallelogram. A *parallelogram* is a four-sided figure with opposite sides parallel. All of the parallelograms below have the same area.

1. Describe how each of the parallelograms **b–e** could be changed into figure **a.**

2. How can your method be used to find the area of any parallelogram?

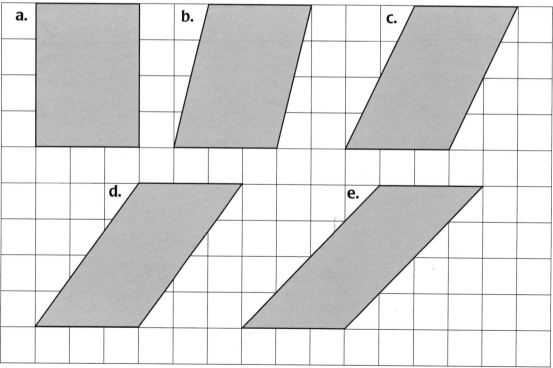

1. Answers will vary. Sample student responses:

   Take the upper, left-hand corner of each parallelogram and shift the shape left.

   Cut a triangular piece from the right side of the parallelogram and paste it onto the left side.

   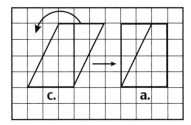

2. Answers will vary. Sample student response:

   You can find the area of any parallelogram by changing it into a rectangle and then counting the number of squares in the figure.

   Any parallelogram with a base measurement and height measurement equal to the length and width of a rectangle will have an area equal to that of the rectangle.

Materials transparency of page 17 of the Student Book, optional (one per class); overhead projector, optional (one per class)

Overview Students are introduced to the terms *quadrilateral* and *parallelogram*. They find a method to determine the areas of parallelograms.

About the Mathematics The areas of most parallelograms can be found using the compensating strategy: cutting and pasting triangular sections of the parallelogram to reshape the figure into a rectangle. Applying this strategy to figure **e** is more difficult; the compensating strategy must be used twice. Note that the strategy of framing the parallelogram within a rectangle and subtracting the remaining parts can always be used.

Planning Discuss the term *quadrilateral* and give several examples. Also discuss the characteristics of a parallelogram. Ask students what they know about parallel lines and how they can determine whether lines are parallel or not. (Two parallel lines will always be equidistant from each other.) Students may work in pairs or in small groups on problems **1** and **2**. Be sure to discuss students' answers to both problems.

Comments about the Problems

1. Some students may say: *You can put the parallelogram straight up.* Be aware that they may possibly think that the height will increase, as indicated below.

   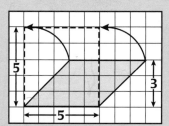

   With this transformation, the area will always be larger than that of the original parallelogram. On page 19 of the Student Book, the idea of *shifting* will be further developed.

2. Students may say that any parallelogram can be changed into a rectangle by cutting and pasting triangular sections.

Not all quadrilaterals are as easy to work with.

Below are some quadrilaterals inside rectangles. Each corner of each quadrilateral touches one side of the rectangle.

3. Calculate the areas of the shaded quadrilaterals.

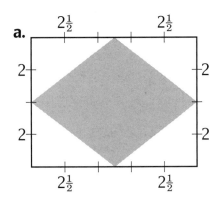

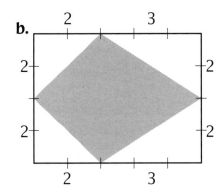

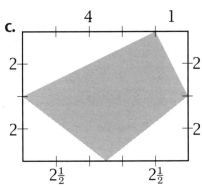

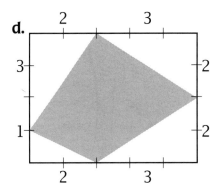

4. Do you think that there is a rule for finding the area of a quadrilateral whose corners touch the sides of a rectangle? If so, explain the rule.

5. Draw some other quadrilaterals on grid paper to see if your rule works. If you do not have a rule, check another student's rule. Try to disprove it.

6. It may be that the rule works on only some of the examples. Describe the situations in which the rule does work.

Britannica Mathematics System

3. a. 10 square units   **b.** 10 square units

    c. 10 square units   **d.** 10 square units

Strategies will vary. Students may subtract the areas of the corner triangles from the area of the rectangle.

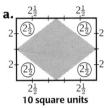 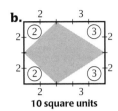

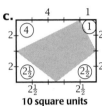 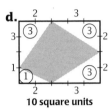

Students may divide the rectangle into smaller rectangles and use triangles to find the area.

4. Answers will vary. There is a rule, which is that a quadrilateral whose sides touch a rectangle and that has at least one diagonal parallel to a side of the rectangle will have $\frac{1}{2}$ the area of the rectangle.

5. Answers will vary. Sample student response:

The rule works for these quadrilaterals:

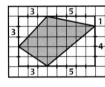

Area of rectangle =
$8 \times 5 = 40$ square units

Area of quadrilateral =
$\frac{1}{2} \times 40$ square units =
20 square units

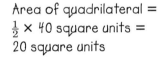

Area of rectangle =
$7 \times 3 = 21$ square units

Area of quadrilateral = $\frac{1}{2} \times 21$ square units = $10\frac{1}{2}$ square units

The rule does not work for this quadrilateral:

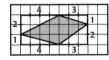

6. Answers will vary. Sample student response:

The rule only works when at least two corners of the quadrilateral are on the same grid line.

Materials grid paper (one or two sheets per student)

Overview Students find the areas of quadrilaterals that are bordered by rectangles. They then develop a rule for finding these areas and investigate whether their rule holds for other quadrilaterals.

Planning Students can work on problem **3** in pairs or in small groups. Problem **3** may be assigned as homework. Be sure to discuss this problem before students continue with problems **4–6.** You might have students work individually on problems **4–6** and use them as assessments.

Comments about the Problems

3. Homework This problem may be assigned as homework. You might want students to create diagrams that show their computations and strategies and share them during a class discussion. You might need to discuss the issue of units for these measurements. For example:

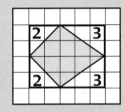   or 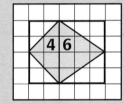

4. Students should see in problem **5** that the rule (the area of the quadrilateral will always be half of the area of the bordering rectangle) is not always true. Any quadrilateral that does not have one diagonal that is parallel to the sides of the rectangle will not have an area that is half of the rectangle's.

4–6. Informal Assessment These problems assess students' ability to generalize formulas and procedures for determining areas of rectangles, triangles, parallelograms, quadrilaterals, and circles.

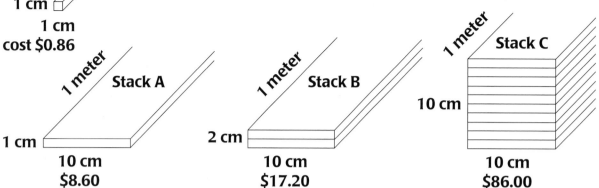

Balsa is a lightweight wood used to make model airplanes. For convenience, balsa is sold in standard lengths. This makes it easy to calculate prices. The price of a 1-by-1-centimeter board that is 1 meter long is 86¢.

1 meter

1 cm

1 cm
1 cm
cost $0.86

Stack A

1 meter
1 cm
10 cm
$8.60

Stack B

1 meter
2 cm
10 cm
$17.20

Stack C

1 meter
10 cm
10 cm
$86.00

7. Explain how the prices for the three stacks shown above were determined.

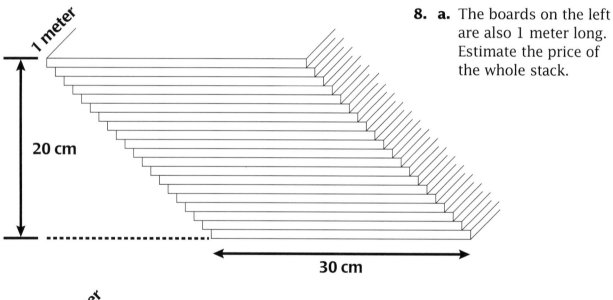

1 meter

20 cm

30 cm

8. a. The boards on the left are also 1 meter long. Estimate the price of the whole stack.

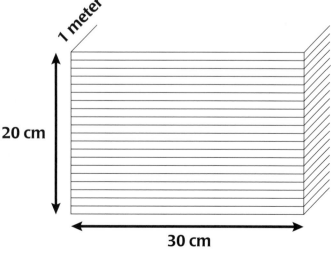

1 meter

20 cm

30 cm

b. The craft shop owner straightened the stack. Now it is much easier to calculate what the stack is worth. Calculate the total price.

c. Compare this with your initial estimate.

7. Answers will vary. Sample student response:

Stack A is made of 10 boards that measure 1 centimeter by 1 centimeter by 1 meter. The price for one board this size is $.0.86, so the price for 10 boards is 10 × $0.86 = $8.60.

Stack B is made of two layers of stack A, so the price is 2 × $8.60 = $17.20.

Stack C is made of five layers of stack B, or 10 layers of stack A, so the price is 5 × $17.20, or 10 × $8.60 = $86.

8. a. Estimates will vary. Accept estimates between $450 and $600.

b. $516 (3 × 2 × $86 = $516) (The stack is three times as wide and twice as tall as stack C from problem **7**.)

c. Answers will depend on students' estimates from problem **8a.**

Materials rulers, optional (several per group)

Overview Students explain how the prices for different stacks of balsa wood were determined. They estimate the price of a slanted stack of balsa wood and calculate the total price for the same stack after it has been straightened.

About the Mathematics This context of straightening a stack of boards is intended to reinforce students' understanding of finding the area of a parallelogram. By comparing the slanted stack with the straightened stack, students may notice that only the width of the boards and the number of boards, or the height, determine the price of the stack. The concept of *shifting* to transform a parallelogram into a rectangle can be used to find a rule to determine the area of a parallelogram. The compensation strategy used in problems **1** and **2** can also be used to devise a rule or formula for finding the area of a parallelogram.

Planning Students can work on problems **7** and **8** individually or in small groups. Discuss these problems, focusing on the idea behind the context of the slanted and straightened stacks, rather than on the computations involved.

Comments about the Problems

8. If the students have difficulty visualizing the context, it may be helpful to demonstrate this idea using a stack of rulers or books.

Did You Know? The wood of the balsa tree is remarkably light and strong and does not bend easily. It is one of the most rapidly growing trees of the tropical forests of Central and South America. The wood is lighter than cork. Ecuador grows most of the world's supply of balsa wood.

In Section A, you learned to reshape figures. You cut off a piece of a shape and pasted that same piece back on in a different spot.

Below are three parallelograms. The first diagram shows how to transform the parallelogram into a rectangle by cutting and pasting.

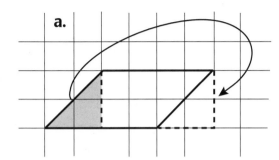 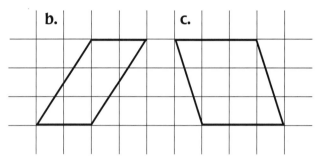

9. Copy the other two parallelograms onto grid paper and show how to transform them into rectangles.

10. Calculate the areas of all three parallelograms.

Activity

Looking for Patterns

You can transform a rectangle into many different parallelograms by cutting and pasting a number of times. Try this on grid paper.

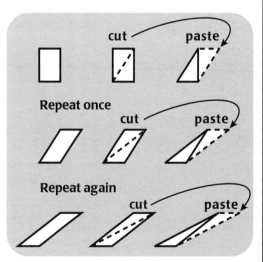

   i. Draw a rectangle that is two units wide and three units high.

  ii. Cut along a diagonal and then paste to create a new parallelogram.

 iii. Repeat step **ii** a few more times.

11. How is the final parallelogram different from the rectangle? How is it the same?

9.

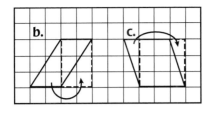

10. **a.** 8 square units (4 × 2 = 8 square units)

   b. 6 square units (2 × 3 = 6 square units)

   c. 9 square units (3 × 3 = 9 square units)

11. Answers will vary. Sample student response:

   The parallelogram is different from the
   rectangle because it has a different shape.
   It looks like the top of the rectangle has been
   shifted to the right to make the parallelogram.
   The figures are the same because they have
   the same base and the same area.

Materials grid paper (two sheets per student);
scissors (one pair per student); glue or tape (one
bottle or roll per pair or group of students);
transparency of grid paper, optional (one per
class); overhead projector, optional (one per class)

Overview Students use the cutting and pasting
strategy to transform parallelograms into
rectangles and vice versa.

About the Mathematics The pictures in the activity
on page 20 of the Student Book show that a diagonal
of a parallelogram divides the parallelogram into
two congruent triangles. Therefore every triangle
can be considered as half of a parallelogram.

This activity illustrates one of the properties of
area: no matter how a shape is rearranged, the
area of the shape remains the same. Students may
not be aware that the height and base of each
figure pictured in the activity also remain the
same. You might want to explain these concepts
and terms using an overhead projector with a
transparency of a piece of grid paper.

Planning Students may work individually on
problem **9** and in pairs or in small groups on
problems **10–11.** Problem **9** is optional. Allow
students to use rulers to draw the diagonals at
each step (as guidelines for cutting) in problem **10.**
Discuss students' responses to problem **10** before
they work on problem **11.**

Comments about the Problems

9. **Informal Assessment** This problem
   assesses students' ability to compare the
   areas of shapes using a variety of strategies
   and measuring units. It is designed to help
   students realize that every parallelogram can
   be transformed into a rectangle with the
   same area. Some students may need to cut
   and paste to develop this connection.

11. Students will probably comment on the fact
    that the shapes formed by cutting and pasting
    look different from the original shape. When
    discussing this problem, ask students to
    explain what they know about the areas of
    the parallelograms. Most will respond that
    the areas are equal. A few students may
    notice that the bases and heights of the
    parallelograms are equal.

Five Square Units

12. a. On grid paper, draw eight different shapes, each with an area of five square units.

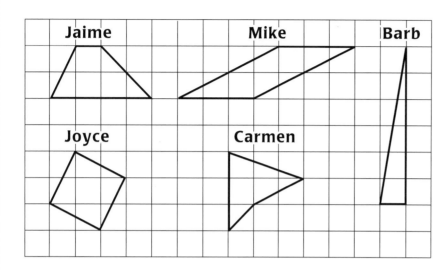

Jaime, Mike, Joyce, Carmen, and Barb drew the shapes on the left.

b. Does each shape have an area of five square units? Explain how you found each area.

c. Draw a triangle that has an area of five square units.

Here are three shapes you have worked with in this section. When you describe a rectangle, parallelogram, or triangle, you can use the words shown here. The *base* describes how wide the figure is. The *height* describes how tall it is.

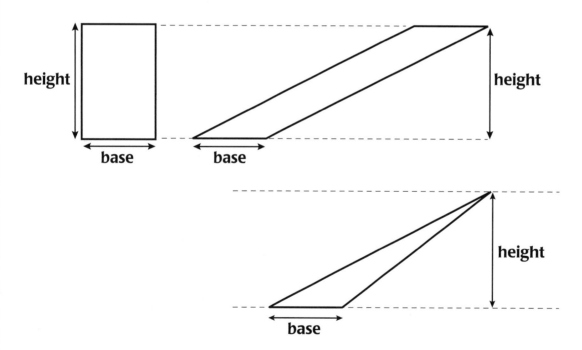

13. Use the words *base* and *height* to describe ways to find the areas of rectangles, parallelograms, and triangles. Be prepared to justify your descriptions.

12. a. Sample student drawing:

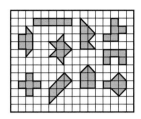

b. No, only Jaime's and Joyce's two shapes have an area of 5 square units. Explanations will vary. Sample student explanations:

Jaime:

Divide the shape into three parts.
Area of part a = $\frac{1}{2} \times 2 \times 1 = 1$ square unit
Area of part b = 2 square units
Area of part c = $\frac{1}{2} \times 2 \times 2 = 2$ square units
Area = $1 + 2 + 2 = 5$ square units

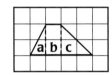

Mike:

Draw a rectangle around it. The area of this rectangle is $7 \times 2 = 14$. The area of the parallelogram is $\frac{1}{2}$ the area of the rectangle, or 7 square units.

Barb:

Draw a rectangle around it. The area of this rectangle is $1 \times 6 = 6$ square units. The area of my shape is half of this area, or 3 square units.

Joyce:

Draw a rectangle around it and find the area: $A = 3 \times 3 = 9$ square units. Then subtract the areas of the triangles:
$A = 9 - 4 = 5$ square units.

Carmen:

Divide it into three triangles and a square.
$A = \frac{1}{2} + 1 + 1\frac{1}{2} + 1 = 4$ square units.

13. Answers will vary. Students' responses should include the following information:

The area of a rectangle is *base* times *height*. The area of a parallelogram is *base* times *height*. The area of a triangle is $\frac{1}{2}$ of *base* times *height*.

Materials rulers, optional (one per student); grid paper (one sheet per student): scissors, optional (one pair per student); glue or tape, optional (one bottle or roll per pair or group of students); transparency of grid paper, optional (one per class); overhead projector, optional (one per class)

Overview Students describe ways to find the areas of rectangles, parallelograms, and triangles. They also draw eight different shapes on grid paper, each of which has an area of five square units.

About the Mathematics It is more important for students to understand the *concept* of area as it relates to rectangles, parallelograms, and triangles than to use rules or formulas for finding the areas of these shapes.

Planning Students can work on problem **12** in pairs or in small groups. Problem **13** may be used as an assessment.

Comments about the Problems

12. a. You may want students to check each other's shapes to see that the area of each figure is five square units. Students do not have to explicitly calculate areas to answer this problem.

13. Informal Assessment This problem assesses students' ability to generalize formulas and procedures for determining the areas of rectangles, triangles, parallelograms, quadrilaterals, and circles.

If students have difficulty seeing that the parallelogram on page 21 of the Student Book has the same area as the rectangle, you may show how the parallelogram can be transformed into the rectangle on the overhead using the same procedure as shown on page 20 of the Student Book.

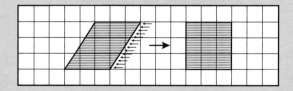

Strategies and Formulas

The area of a parallelogram is the same as the area of a rectangle with the same base and height. In other words, the area of a parallelogram is found with this formula: The area (A) is equal to the base (b) times the height (h).

$$A_{\text{rectangle}} = A_{\text{parallelogram}} = b \times h$$

A triangle is always half of a parallelogram. So the area of a triangle can be found with this formula: Area is equal to one-half of the base times the height.

$$A_{\text{triangle}} = \tfrac{1}{2}b \times h$$

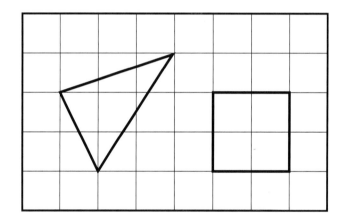

It is not always easy to establish the height and the base of a triangle. Sometimes you have to use some other strategy to find the area of a triangle. This is true for the triangle on the right.

14. The triangle and the square on the right are cut out of the same material.

   a. The square weighs 2 pounds. How much does the triangle weigh?

   b. Elissa solved this problem by cutting and pasting four small pieces (see the diagram on the right). Redo her work using a piece of grid paper. What answer do you think Elissa found?

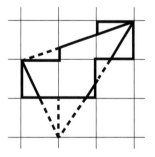

   c. Miguel drew a square and subtracted the areas of three triangles (see the diagram on the right). Describe his strategy and calculate his answer.

   d. Are the answers that Elissa and Miguel found the same?

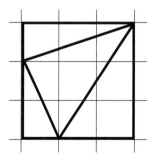

14. a. $1\frac{3}{4}$ pounds. (The square contains 4 units, each weighing $\frac{1}{2}$ pound. The triangle contains $3\frac{1}{2}$ square units, so its weight is $3\frac{1}{2} \times \frac{1}{2}$, or $1\frac{3}{4}$ pounds.)

b. Elissa found that the weight was $1\frac{3}{4}$ pounds because the area is $3\frac{1}{2}$ square units.

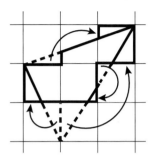

c. Answers will vary. Sample student response:

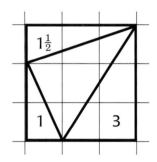

First he found the area of the square.
A = 3 × 3 = 9 square units

Then he found the areas of the triangles.
$\frac{1}{2}$ × 2 × 1 = 1 square unit
$\frac{1}{2}$ × 2 × 3 = 3 square units
$\frac{1}{2}$ × 1 × 3 = $1\frac{1}{2}$ square units

Then he added the areas of the triangles.
1 + 3 + $1\frac{1}{2}$ = $5\frac{1}{2}$ square units

Finally he subtracted the areas of the triangles from the area of the square.
A = 9 − $5\frac{1}{2}$ = $3\frac{1}{2}$ square units

d. Yes.

Materials grid paper (one sheet per student); scissors (one pair per student); glue or tape (one bottle or roll per group); transparency of grid paper, optional (one per class); overhead projector, optional (one per class)

Overview Students are introduced to the formulas for finding the area of a parallelogram, a rectangle, and a triangle. They see how the strategies of cutting and pasting and of subtraction are used to find the area of a triangle.

Planning Have a short class discussion about the two area formulas introduced here. Students can work on Problem **14** individually or in pairs. This problem can be used as an assessment and/or assigned as homework.

Comments about the Problems

14. Informal Assessment This problem assesses students' ability to estimate and compute the areas of geometric figures; to compare the areas of shapes using a variety of strategies and measuring units; to understand which units and tools are appropriate to estimate and measure area, perimeter, and volume; and to use the concepts of perimeter, area, and volume to solve realistic problems.

This is the first problem that requires students to show that they can apply various strategies. Encourage students to show their calculations. You might have them illustrate the strategies and their calculations using a transparency of a piece of grid paper.

c. Students can write the area of each triangle to be subtracted on their drawings. Some students will be confused when they draw rectangles around the outside triangles because the rectangles overlap:

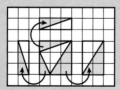

You may suggest that students draw the figure three times in order to focus on only one triangle in each drawing.

Smart Counting

Here is a drawing of a berry bush with many branches. In this drawing, the branches and berries form a regular pattern.

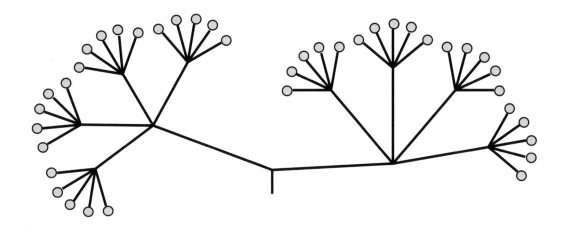

15. **a.** How many berries are there?

   b. How did you calculate the number of berries?

Below is a tiled terrace.

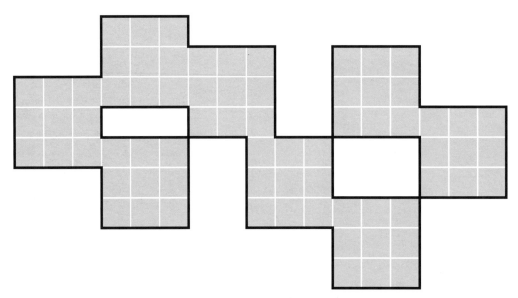

16. **a.** Can you see a pattern in the terrace? What is the pattern?

   b. Describe a way to calculate the number of tiles in the terrace without counting each individual tile.

15. a. 40 berries

    b. Methods may vary. Sample student response:

> There are eight branches, and each one
> has five berries, so I multiplied
> $8 \times 5 = 40$ berries.

16. a. Answers will vary. Sample student response:

> The terrace is made of groups of tiles, and
> each of these groups has the same number of
> tiles. There are eight large squares and each
> square is made up of nine small squares.

    b. Answers will vary. Sample student responses:

> To calculate the number of tiles in the terrace
> without counting, look for groups of tiles.
> This terrace has eight large squares, and
> each square has nine individual tiles. The
> terrace has $8 \times 9 = 72$ tiles.

> The terrace has 24 rows of tiles and
> each row has three tiles. There are
> $24 \times 3 = 72$ tiles on the terrace.

> This terrace has two groups of four large
> squares. Each square has nine tiles. The
> total number of tiles in this terrace is
> $2 \times 4 \times 9 = 72$.

Overview Students use patterns to count berries
in an illustration of a berry bush and tiles in a
pictured terrace.

About the Mathematics Solving these problems
requires students to apply their understanding of
area as well as use smart counting methods.
Counting strategies can be used even when
shapes other than square units are used to
measure area.

Planning Students may work on problem **15** in
pairs or in small groups. Problem **15** can be
assigned as homework. Problem **16** can be used
as an assessment. Students' strategies for these
problems can be discussed in the next class
session. Accept all accurate methods of calculating
the totals.

Comments about the Problems

15. Homework This problem may be assigned
as homework. Encourage students to use a
method other than counting the berries one
by one.

16. Informal Assessment This problem
assesses students' ability to represent and
solve problems using geometric models.

On the right is the plan for the main walkway in a new mall. The floor has been made from small, triangular tiles. Use the drawing of the floor on **Student Activity Sheet 9** to solve problem **17.**

A Small Tile

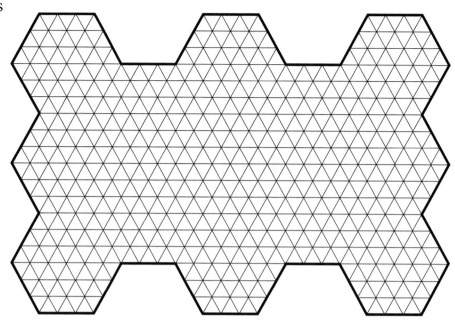

17. How many small tiles were used to create the floor? (Think of a fast way to count the tiles.)

Elissa wants to make the floor on the right using large sections of tile like the one on the right. The tiles can be cut apart.

A Large Section of Tile

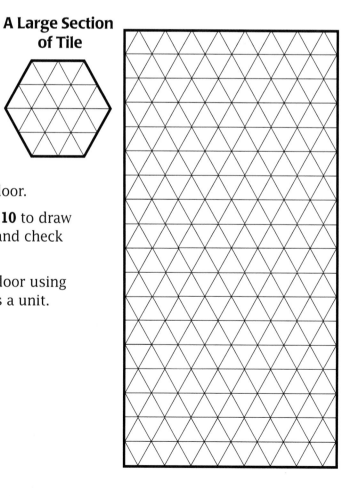

18. a. Estimate, without drawing, how many large sections of tile are needed to cover the floor.

   b. Use **Student Activity Sheet 10** to draw the large sections of tile and check your estimate.

   c. Describe the area of the floor using the large section of tile as a unit.

17. 702. Strategies will vary. Sample student strategy:

There are 13 large hexagons. Each hexagon is made of six large triangles, and each triangle is made of nine small tiles: $13 \times 6 \times 9 = 702$ small tiles.

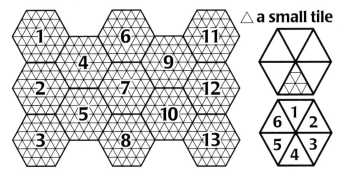

△ **a small tile**

18. a. Estimates will vary. Accept estimates between 10 and 14 sections of tiles.

b. Drawings will vary. Sample student drawings:

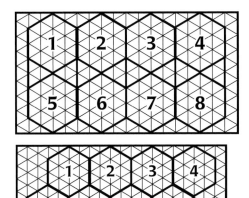

c. Answers will vary. Sample student response:

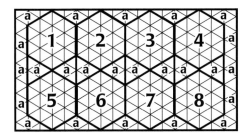

The area of the floor is 12 hexagonal sections of tile. Eight hexagons fit on the floor without cutting. There are 24 empty spaces, which I labeled a, and 4 tiles fit into each of these spaces. (A few tiles will need to be cut in half.) Six section a's will fit into one hexagon, so 24 section a's are four more hexagons. The total area is $8 + 4 = 12$ hexagons.

Materials Student Activity Sheets 9 and 10 (one of each per student); transparency of Student Activity Sheet 9, optional (one per group of students); overhead projector, optional (one per class); scissors, optional (one pair per student); glue or tape, optional (one bottle or roll per pair or group of students)

Overview Students develop a strategy for counting the number of triangular tiles in a large floor. They estimate and then calculate the number of large hexagonal sections of tile that are needed to cover the area of a floor.

About the Mathematics Problem **18** revisits the concept of area. Students can first count the number of whole hexagons that fit in the floor space and then combine the remaining floor areas to determine the total number of hexagons needed. Another strategy is to use a refinement of the measurement unit. In this problem, the refinement could be a small triangle. Twenty-four small triangles can be exchanged for one hexagon.

Planning You may want to distribute one transparency of Student Activity Sheet 9 to each small group so that students can show their strategies to the whole class. They may work on problem **17** in pairs or in small groups. This problem may also be assigned as homework. Problem **18** may be used as an assessment.

Comments about the Problems

17. Homework This problem may be assigned as homework. Encourage students to use a method other than counting the individual tiles. Some students may have difficulty seeing the hexagons in the floor pattern and may use other strategies.

18. Informal Assessment This problem assesses students' ability to create and work with tessellation patterns; to understand which units and tools are appropriate to estimate and measure area, perimeter, and volume; and to use the concepts of perimeter, area, and volume to solve realistic problems.

Encourage students to actually cut out the shapes if they need to. You may need to remind some students to use the strategy of cutting and pasting the leftover sections.

Square Patterns

The previous problems show that it can be helpful to group smaller units to make larger units when counting. It is sometimes helpful to exchange units in area problems, too. Here are two shapes with different sizes of square measuring units.

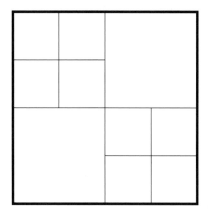 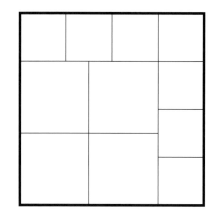

19. How many different-sized measuring units do you see in the two shapes above?

On **Student Activity Sheet 11** label the smallest square **A,** the next larger size **B,** the next larger size **C,** and so on.

20. How many squares of size **A** does it take to make a square of size **C?**

21. How many squares of size **C** does it take to make a square of size **E?**

22. How many squares of size **A** does it take to make a square of size **E?**

Square Shifts

Look again at the two squares above. The square on the left is filled with 10 squares. The square on the right is filled with 11 squares. Use the empty squares on **Student Activity Sheet 11** to solve the following problem.

23. It is possible to fill a square with 12 smaller squares? What about 13 squares, or 14? The squares can be of different sizes but together must fit the large square exactly. Try to find out what possibilities there are.

Britannica Mathematics System

19. There are three different-sized squares shown in the shape on the left and four different-sized squares shown in the shape on the right.

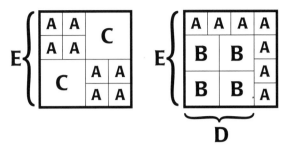

D is made of four **B**'s, and **E** is the entire square.

20. Four squares of size **A** will fit into one square of size **C.**

21. Four squares of size **C** will fit into one square of size **E.**

22. Sixteen squares of size **A** will fit into one square of size **E.**

23. Yes, 12, 13, or 14 smaller squares will fit into one larger square.

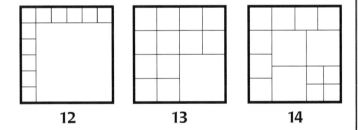

| 12 | 13 | 14 |

Materials Student Activity Sheet 11 (one per student); colored pencils, optional (one box per group of students)

Overview Students find all the different-sized square measuring units in an illustration and investigate how they are related. They also investigate the number of smaller squares into which a single square can be divided.

Planning Make enough copies of Student Activity Sheet 11 so students who are having difficulty can cut and paste to answer the questions. Problem **23** is optional. Some students may find this problem challenging. Problems **19–23** may also be assigned as homework. Students can work on these problems in pairs or in small groups.

Comments about the Problems

19. Homework This problem may be assigned as homework. Some students may find it helpful to use colored pencils to indicate the different-sized measuring units. Students should focus on only square units.

23. Homework This problem may be assigned as homework. This investigation digresses somewhat from the focus of the unit. However, students may find these mathematical brain teasers interesting.

Extension Ask students to find a relationship between the lengths of the sides of square **B** and square **C** and write their response in their journals. (See the solutions column.)

Summary

In this section, you learned many ways to calculate the areas of a variety of shapes. To find the area of the shape on the right, you can use any of the following strategies:

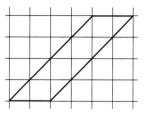

Count, Cut, and Paste Partial Units

Count the number of complete squares inside the shape, cut out the remaining pieces and move them to form new squares.

Step 1

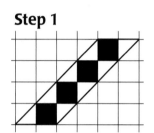

Step 2

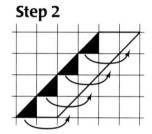

Reshape the Figure

Cut off larger parts of the original figure and paste them somewhere else.

Step 1

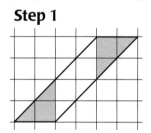

Step 2

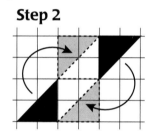

Enclose the Shape and Subtract Extras

Draw a rectangle around the shape in such a way that you can easily subtract the areas of the rectangle that are not part of the shape.

In this case, the area of the parallelogram is the area of the rectangle minus the areas of the two triangles.

$24 - 8 - 8 = 24 - 16 = 8$ square units

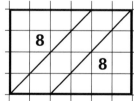

Double the Shape or Cut It in Half

The area of the shaded triangle is half the area of the corresponding rectangle.

Overview This page summarizes the different strategies students have used to find the area of shapes. These strategies include:

• counting the whole units and cutting and pasting partial units to make additional whole units,

• reshaping the figure,

• enclosing the shape and subtracting the extra parts, and

• doubling or halving known areas.

Planning You may ask students to read the Summary aloud and then discuss the strategies in class. There are no problems for students to solve on this page.

Extension Ask students to design other shapes that could be exchanged for the shapes described on Student Book page 26. Students can then show how the four strategies can be used to determine the areas of the shapes they designed. Allow students to draw different shapes for each strategy if they want to.

Summary, continued

In each of the examples on page 26, a parallelogram with a base of 2 and a height of 4 had an area of 8 square units. You can use the following rule:

> The area of a parallelogram is equal to the area of a rectangle with the same base and the same height.

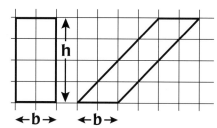

This rule gives you the following formula:
The area of a parallelogram is equal to the length of the base times the height.

$$A_{\text{parallelogram}} = b \times h$$

In this section, you calculated the area of shapes using units of measure that were not squares. Areas were expressed using the number of small, triangular tiles or the number of larger, hexagonal sections of tile needed to fill a shape.

Summary Questions

24. a. Copy the parallelogram on the right and shade a rectangle that is the same area.

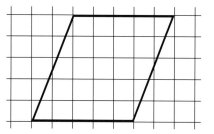

b. Use your copied parallelogram to find a triangle whose area is half the area of the parallelogram.

24. a. Drawings will vary. Sample student drawing:

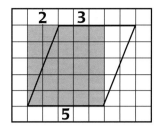

b. Drawings will vary. Sample student drawings:

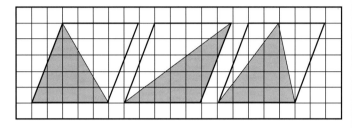

Area of parallelogram = 5 × 5 = 25 square units
Area of a triangle = $\frac{1}{2}$ × 5 × 5 = $12\frac{1}{2}$ square units

Materials transparency of Figure 1 with a grid overlay, optional (one per class); overhead projector, optional (one per class)

Overview Students read the continuation of the Summary on this page. The area of the parallelogram on the previous page of the Student Book is now related to its base and height. The formula for the area of a parallelogram is also reviewed. Students find a rectangle that has the same area as a given parallelogram and a triangle whose area is exactly half that of the parallelogram.

Planning Be sure to discuss the main concepts outlined in the Summary. Before students begin problem **24,** you may want to display a transparency of a grid overlaid with a copy of a parallelogram (see Figure 1 below) on the overhead projector and ask students how to find the area of the parallelogram. Then turn the parallelogram so another side of the parallelogram is the base, and ask students how to find the area of the parallelogram in this position (see Figure 2 below).

Figure 1

Figure 2

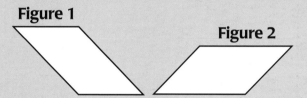

Students may work in pairs or small groups on problem **24.** Discuss problem **24** with the whole class.

Comments about the Problems

24. a. If students are having difficulty, suggest that they look at page 20 of the Student Book, where a rectangle is transformed into a parallelogram. Problem **24** can be solved by working backwards. Some students may determine the base (5 units) and height (5 units) and understand that any rectangle with a base of 5 units and a height of 5 units will have an area of 25 square units.

Summary Questions, continued

25. Determine the areas of the following shapes. Use any method.

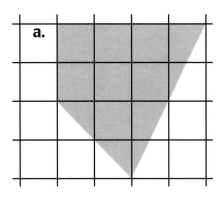

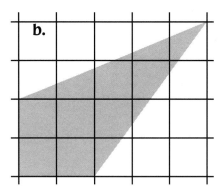

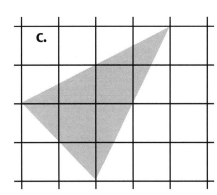

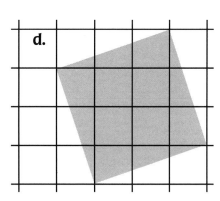

26. A bathroom floor can be covered with 600 small tiles. How many tiles will be needed for the same bathroom floor if the larger tiles are used instead of the small ones?

25. a. 10 square units

b. 9 square units

c. 6 square units

d. 10 square units

25. 150 larger tiles. (Since four small tiles cover the same amount of space as one large tile, one-fourth of the number of smaller tiles is needed.)

Materials grid paper, optional (one sheet per student); scissors, optional (one pair per student); glue or tape, optional (one bottle or roll per pair or group of students)

Overview Students determine the areas of different figures and solve a problem about small and large floor tiles.

About the Mathematics Problem **26** shows the relationship between the factors of enlargement of sides and of area: the side of the large tile is two times larger than the side of the small tile, while the corresponding area of the large tile is four times greater than that of the small tile. If students gain this insight, they can develop a better understanding for exchanging area measuring units, one of the topics of Section D.

Planning You may want students to work on problem **25** individually and use it as an informal assessment. Students may work in pairs or in small groups on problem **26.** Have students share their strategies. After students complete Section C, you may assign appropriate activities from the Try This! section, located on pages 49–52 of the Student Book, as homework.

Comments about the Problems

25. Informal Assessment This problem assesses students' ability to estimate and compute the areas of geometric figures and to understand which units and tools are appropriate to estimate and measure area, perimeter, and volume.

26. Have students explain their answers. Ask them to mention all the relationships they see. For example, the area of the large tile is four times the area of the small tile, the area of the small tile is one-fourth the area of the large tile, and a side of the small tile is one-half the length of a side of the large tile.

Work Students Do

Students find familiar objects that are one centimeter, one meter, and one kilometer in length. They measure the areas of various regions using square centimeters, square meters, square inches, and square yards and look at the relationships among these units of measure. They also study the relationships among square yards, square feet, and square inches. Finally, they solve realistic problems involving floor coverings and population density.

Goals

Students will:

- understand which units and tools are appropriate to estimate and measure area, perimeter, and volume;
- understand the structure and use of standard systems of measurement, both metric and English;
- use the concepts of perimeter, area, and volume to solve realistic problems.

Pacing

- approximately four 45-minute class sessions

Vocabulary

- kilometer
- meter

About the Mathematics

In this section, students use metric and English units of measure. The difference between a linear unit of measure and a square unit of measure is addressed. The metric system is based on refinements in tenths. One millimeter is a refinement of one centimeter; it is one-tenth of a centimeter. Refinements of units of measure is one of the subjects in the grade 5/6 unit *Measure for Measure*. Studying metric units helps students develop an understanding of decimal numbers.

The relationships between square units of different sizes can be found by calculating how many smaller units fit into each larger unit. For example, one square centimeter has 10 rows of 10 square millimeters; therefore, a total of 100 square millimeters fit inside it.

In the same way, the relationships between English square units of measure can be found: in one square yard, three rows of three square feet will fit; so one square yard is nine square feet. Students do not convert between the metric and English systems.

Materials

- centimeter rulers, pages 69 and 71 of the Teacher Guide (one per student)
- meter sticks, pages 69 and 77 of the Teacher Guide (one per group of students)
- square pieces of paper with sides of 1 meter, pages 71 and 75 of the Teacher Guide, optional (one per class)
- square pieces of paper with sides of 1 yard, page 71 of the Teacher Guide, optional (one per class)
- inch rulers, page 71 of the Teacher Guide, (one per student)
- millimeter grid paper, page 75 of the Teacher Guide, optional (one sheet per pair or group of students)
- typing paper, page 77 of the Teacher Guide, optional (one ream per class)
- drawing paper, page 77 of the Teacher Guide, optional (10 sheets per group of students)
- transparency of a square with sides of 10 centimeters, page 77 of the Teacher Guide, optional (one per class)
- pennies, page 77 of the Teacher Guide, optional (one per student)
- glass container filled with dry beans or jelly beans, page 81 of the Teacher Guide, optional (one per class)

Planning Instruction

In this section, students use metric and English units of measure. If students are not familiar with metric units, you might have some brief class discussions in which you demonstrate how to use centimeter rulers and meter sticks. You can also compare the relative sizes of metric areas by showing students pieces of paper with areas of 1 square centimeter and 1 square meter. Several topics for discussion and activities are suggested throughout this section in the Hints and Comments column. Problems 11–13 might lead into a discussion about volume. The topic of volume is formally introduced in Section E.

Students may work individually on problems 7 and 17–19 and solve problems 4–6 as a class activity or in small groups. They may work on the remaining problems in pairs or in small groups.

Problems 13 is optional. If time is a concern, you may omit this problem or assign it as homework.

Homework

Problems 14 (page 76 of the Teacher Guide), 17a (page 80 of the Teacher Guide), and 19 (page 82 of the Teacher Guide) can be assigned as homework. Also, the Bringing Math Home activity (page 69 of the Teacher Guide), the Extension (page 75 of the Teacher Guide), and the Writing Opportunity (page 81 of the Teacher Guide) can be assigned as homework. After students complete Section D, you may assign appropriate activities from the Try This! section, located on pages 49–52 of the Student Book. The Try This! activities reinforce the key mathematical concepts introduced in this section.

Planning Assessment

- Problem 7 can be used to informally assess students' ability to understand the structure and use of standard systems of measurement, both metric and English.
- Problem 12 can be used to informally assess students' ability to use the concepts of perimeter, area, and volume to solve realistic problems.
- Problem 15 can be used to informally assess students' understanding of which units and tools are appropriate to estimate and measure area, perimeter, and volume and their ability to use the concepts of perimeter, area, and volume to solve realistic problems.
- Problem 18 can be used to informally assess students' ability to understand the structure and use of standard systems of measurement, both metric and English, and to understand which units and tools are appropriate to estimate and measure area, perimeter, and volume.

Going Metric

Many people find English units of measurement more difficult to use than metric units, which are based on multiples of 10. The United States is one of the few countries that still use the English system of measurement. But many products in your grocery store (bottled water as well as certain canned fruits and vegetables from Mexico and South America) are measured in metric units. International games like those at the Olympics use metric distances. Medicines are weighed in metric units. Food labels usually list fat, protein, and carbohydrates in metric units.

The descriptions below will help you understand the sizes of some commonly used metric units.

Length

1 centimeter: Your fingernail is about 1 centimeter wide.
1 meter: One giant-step is about 1 meter long.
1 kilometer: Seven city blocks are about 1 kilometer long.

Try to become familiar with these units.

1. Make a list of things that are approximately the size of:

   a. a centimeter

   b. a meter

   c. a kilometer

2. **a.** How many centimeters are in a meter?

   b. How many meters are in a kilometer?

3. Give an example of something that is about the size of:

   a. a square centimeter

   b. a square meter

   c. a square kilometer

1. a. Answers will vary. Sample student responses:

- the width of a paper clip
- the width of a pen

b. Answers will vary. Sample student responses:

- the width of a door
- four dictionaries lined up lengthwise

c. Answers will vary. Sample student responses:

- the distance between my house and my friend's house
- the length of the fence around the playground
- a 100-story building

2. a. 100 centimeters

b. 1,000 meters

3. a. Answers will vary. Sample student responses:

- a child's thumbnail
- a button on a telephone
- the head of a thumbtack

b. Answers will vary. Sample student responses:

- a tabletop
- an unfolded road map
- two bed pillows

c. Answers will vary. Sample student responses:

- 200 football fields
- the land needed for a large shopping mall (including parking lots)

Materials centimeter rulers (one per student); meter sticks (one per group of students)

Overview Students find points of reference for one centimeter, one meter, one kilometer, one square centimeter, one square meter, and one square kilometer.

About the Mathematics The development of points of reference for units of measurement is very important for students, whether the units are English or metric.

Every student should have an idea about the relative sizes of one centimeter, one meter, and so on, in order to estimate the sizes of objects and to convert between metric units. For example, if students have points of reference for one meter and one centimeter, they can estimate that there are 100 centimeters in one meter. Note that in this unit, students do not convert between the two measurement systems.

Planning You might begin this section with a discussion of the metric system. To demonstrate the length of one meter, have students extend their arms until the distance between their hands is about one meter or hold one hand at a distance of one meter from the floor. To show the size of one centimeter, have students examine the width of one of their fingernails. Then discuss other points of reference, for example:

- the height of a door is about two meters;
- the area of a door is about 1.5 square meters;
- a city block is about 150 meters long;
- a city block has an area of about 22,500 square meters. It would take about 45–50 blocks to make a square kilometer.

Students may work in pairs or in small groups on problems **1–3**.

Comments about the Problems

1–3. Students' responses will demonstrate their prior knowledge about the relative sizes of different metric units. Students might not have enough experience to answer questions related to kilometers and square kilometers.

Bringing Math Home Have students make a list of common household products that are labeled in metric units.

Some things that are 1 square meter in area are not square. Think of the amount of paint that will cover 1 square meter. It does not matter what shape you cover with that paint; its area will always be one square meter. In a similar way, each of the figures below encloses 1 square centimeter.

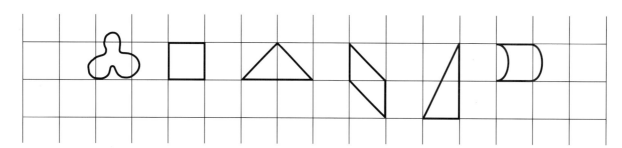

Square units, such as square meters, square centimeters, square yards, and square feet are generally used to measure area. Sometimes it is necessary to compare smaller square units of measure to larger square units of measure.

100 cm

1 m = 100 cm

100 cm

On the left is a square figure with sides 1 meter long. (Note that the figure is not drawn to scale.)

4. a. What is the area of the figure in square centimeters?

b. What is its area in square meters?

A diagram can also be used to compare square inches and square yards. Below is a square with sides 1 yard long; it is not drawn to scale.

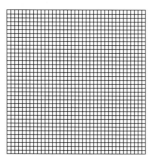

36 inches

1 yard = 36 inches

36 inches

5. a. What is the area of the figure in square inches?

b. What is its area in square yards?

6. Fill in the blanks for the following equations:

    a. 1 square meter = _____ square centimeters = _____ square millimeters

    b. 1 square yard = _____ square feet = _____ square inches

4. a. 10,000 cm^2 (100 cm × 100 cm = 10,000 cm^2)

b. 1 m^2 (1 m × 1 m = 1 m^2)

5. a. 1,296 in.^{2} (36 in. × 36 in. = 1,296 in^2)

b. 1 yd^2 (1 yd × 1 yd = 1 yd^2)

6. a. 1 square meter = 10,000 square centimeters = 1,000,000 square millimeters

b. 1 square yard = 9 square feet = 1,296 square inches

Materials square pieces of paper—one with sides of one meter and one with sides of 1 yard, optional (one of each per class); centimeter rulers (one per student); inch rulers (one per student)

Overview Students investigate and find relationships among metric and English square units of measure.

About the Mathematics The relationships among square yards, square feet, and square inches are further investigated in problem **7** on page 31 of the Student Book. Students explore the relationship between square meters and square centimeters in problem **8** on page 32 of the Student Book.

Planning You might read and discuss the text at the top of Student Book page 30 as a class activity. Have students draw 1-centimeter squares and then design three shapes with areas of 1 square centimeter each. Students may work on problems **4–6** in small groups or as a class. Have students explain their answers to problem **6.**

Comments about the Problems

4–5. The illustrations on Student Book page 30 are not drawn to scale. To give students an idea about the actual size of one square meter in problem **4,** you might do the following activity:

- Show students a piece of paper with sides of 1 meter. This piece of paper shows the actual size of one square meter.

- Ask students how many square centimeters are needed to cover the paper and how they can find this number. [Students may say: *You need 100 rows of 100 square centimeters, or 10,000 square centimeters.*]

You might show students a piece of paper with sides of 1 yard, as in problem **5,** and ask them similar questions. Have students first draw a square with sides of 1 inch.

6. If students are having difficulty, have them draw squares with areas of 1 square centimeter and others with areas of 1 square millimeter. Have them compare the two areas and find the relationship between the two square units. Students can use the same strategy to find the relationship between 1 square foot and 1 square inch.

Floor Covering

A lobby of a new hotel is 14 yards long and 6 yards wide. It needs some type of floor covering.

6 yards

14 yards

There are four options:

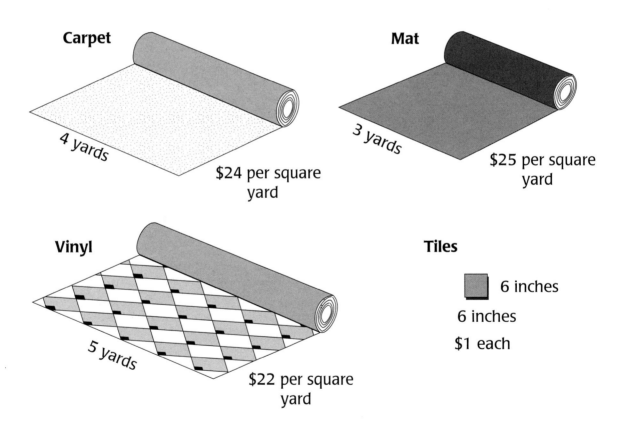

Carpet

4 yards

$24 per square yard

Mat

3 yards

$25 per square yard

Vinyl

5 yards

$22 per square yard

Tiles

6 inches

6 inches

$1 each

7. You are the salesperson for the floor-covering company. You have been asked by the hotel manager to present the total cost of each option along with your recommendation and your reasons for it. Write a report that explains your choice for the best floor covering for the hotel lobby.

7. Answers will vary. Sample student response:

The area of the lobby is 84 square yards
(14 × 6 = 84 yd²). Using carpet will cost $2,016
(84 × $24 per yd²). Two ways to lay the carpet
are shown in this diagram. You can see there's
no waste.

Carpet

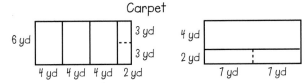

Using a mat will cost $2,100 (84 × $25 per yd²). Two
ways to place the mat are shown in this diagram.
There's no waste using this material either.

Mat

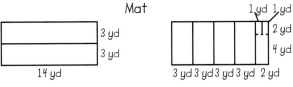

Here are two ways to use vinyl flooring. If you
choose the way shown on the left below, you
need 3 strips (5 yards wide and 6 yards long),
or 90 square yards. This way costs $1,980
(90 × $22 per yd²), and 6 square yards are wasted.
Using the way shown on the right below, you
need 1 strip 5 yards wide and 14 yards long, or
70 square yards. You also need 3 strips (5 yards
long and 1 yard wide), or 15 square yards. The
total for this floor covering is 85 square yards,
so the cost is $1,870 (85 × 22 per yd²), and
one yard is wasted.

Vinyl

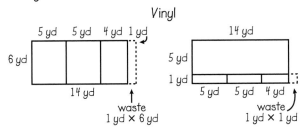

Tiles are the most expensive choice, but they
are easy to clean and last a long time. You need
36 rows by 84 rows of tiles, or 3,024 tiles. Since
1 tile costs $1, the total cost is $3,024.

Tiles

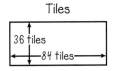

Overview Students investigate four options for
covering the floor of a lobby. They also find the
total cost of each floor covering.

Planning You may want students to work on
problem **7** individually and use it as an
assessment. Briefly discuss students' answers.

Comments about the Problems

7. Informal Assessment This problem
assesses students' ability to understand the
structure and use of standard systems of
measurement, both metric and English.

Buying floor covering involves making
decisions about how much of each covering
must be purchased (depending on the width
of the floor covering and the area of the
room). Students must also consider the costs
of the floor coverings.

You may want to emphasize different
strategies for finding the total costs. For
example, students can compute the price of
the carpet per yard length (4 × $24 = $96),
find out how many yard lengths are needed
(6 + 6 + 6 + 3 = 21 yards), and then
compute the total cost (21 × $96 = $2,016).
Another strategy is to find the number of
square yards that are needed and compute
the total costs (84 × $24 = $2,016). This
strategy will not work if students have
covered the floor in such a way that there is
some leftover floor covering. In this case, the
cost of the leftover covering (such as when
using the vinyl floor covering) must be
included in the total cost.

Some students may have difficulty
converting inches to yards to calculate the
number of tiles needed to cover the floor.
You may suggest that students draw a row
of tiles until it reaches a length of one yard,
using the following line of reasoning:

• a side of one tile is 6 inches;
• a row of two tiles is 12 inches or 1 foot;
• a row of six tiles is 3 feet or 1 yard long.
  Since the lobby is 6 yards wide, they will
  need 36 tiles for each row. Since the lobby
  is 14 yards long, they will need 84 rows
  of tiles.

Several floors of the hotel will be covered with Italian marble. The marble comes in tiles of 1 square meter, which will be carefully cut. The leftover pieces will not be wasted; they can be used in other places where a whole tile will not fit. The dimensions of four floors are shown below.

a.

4 m

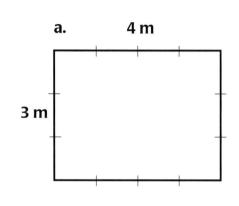

3 m

b.

4 m

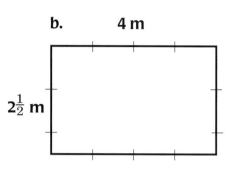

$2\frac{1}{2}$ m

c. $2\frac{1}{2}$ m

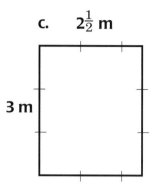

3 m

d. $3\frac{1}{2}$ m

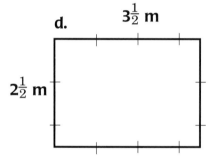

$2\frac{1}{2}$ m

8. How many square meters of marble will be needed for each of these floors?

a. 4 m

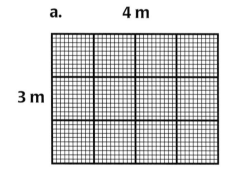

3 m

b. 4 m

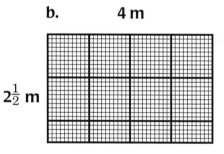

$2\frac{1}{2}$ m

c. $2\frac{1}{2}$ m

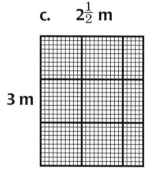

3 m

d. $3\frac{1}{2}$ m

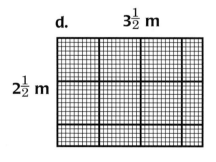

$2\frac{1}{2}$ m

9. Another type of marble tile is available in smaller squares with 10-centimeter sides. How many tiles would be required for each of floors **a** through **d**?

10. How do the answers to problem **9** compare with the answers to problem **8**?

8. a. 12 m^2

b. 10 m^2

c. $7\frac{1}{2} \text{ m}^2$

d. $8\frac{3}{4} \text{ m}^2$

9. a. 1,200 tiles

b. 1,000 tiles

c. 750 tiles

d. 875 tiles

10. The answers to problem **9** are 100 times the answers to problem **8.**

Materials millimeter grid paper, optional (one sheet per pair or group of students); square piece of paper with sides of one meter, optional (one per class)

Overview Students determine the number of 1–square meter and 10-centimeter marble tiles needed to cover four floors.

About the Mathematics Some students may multiply the dimensions of the floors (some of which involve fractions) to calculate the areas. Other students may simply count the numbers of squares to find the areas. Using the area of a rectangle as a model to multiply fractions is made explicit in the grade 7/8 unit *Cereal Numbers*.

The area $(2\frac{1}{2} \times 3\frac{1}{2})$ can be calculated as follows:

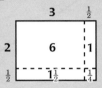

Area $= 6 + 1 + 1\frac{1}{2} + \frac{1}{4} = 8\frac{3}{4}$ square units

Planning Students may work in pairs or in small groups on problems **8–10.** Discuss students' answers to problem **10.**

Comments about the Problems

9. Different strategies can be used to find the number of tiles needed. Some students may count the number of tiles in one row and then count the number of rows. In the first floor there are 40 tiles in each of 30 rows; $40 \times 30 = 1,200$ tiles. Other students may count the number of tiles in a 1–square meter section and then determine how many square meters make up the area of the floor. In the first floor, there are 100 tiles per square meter and there are 12 square meters; so $100 \times 12 = 1,200$ tiles are needed.

10. If students are having difficulty understanding these measurement units, you might want to again show the square with sides of 1 meter and the actual size of a small tile. A few students may know that the area of the small tile is one square decimeter.

Extension Have students do the same activity as in problem **9,** using millimeter grid paper to help them develop strategies for converting between the two units of measure, the square centimeter and square millimeter.

Thick and Thin

A ream of typing paper contains 500 sheets. The ream is about 5 centimeters thick.

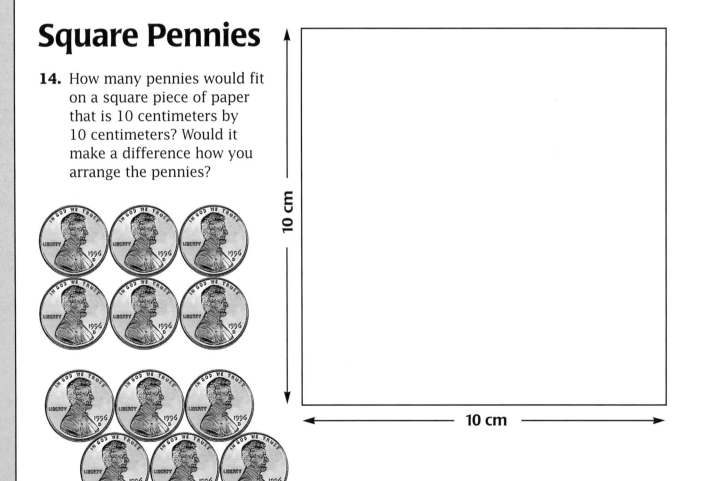

5 cm

1 ream
500 sheets

11. About how thick is one sheet of paper?

12. Would the 500 sheets cover the floor of your classroom? Explain your reasoning.

13. Think of other situations in which something is used to completely cover something else. Describe them in your notebook or journal.

Square Pennies

14. How many pennies would fit on a square piece of paper that is 10 centimeters by 10 centimeters? Would it make a difference how you arrange the pennies?

10 cm

10 cm

11. about one one-hundredth ($\frac{1}{100}$) of a centimeter

12. Answers will vary, depending on the size of your classroom. Sample student response:

No, the paper will not cover our classroom floor. Our classroom is 6 m × 8 m, or 48 m². One piece of paper is 21.5 cm × 28 cm, or about 600 cm². That's the same as 0.06 m². Five hundred sheets cover 500 × .06 m², or 30 m², which is a smaller area than the area of our classroom floor.

13. Answers will vary. Sample student responses:

- covering the gym floor with tumbling mats,
- covering a dining room table with a tablecloth that just fits; it doesn't hang over the sides,
- wallpapering the walls,
- carpeting the floors,
- laying sod on a lawn.

14. Answers will vary. Sample student response:

Up to 28$\frac{3}{4}$, depending on how they are placed and whether you cut them. Yes, the arrangement is important. Here are two ways to place the pennies.

25 pennies

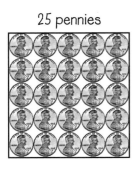

28$\frac{3}{4}$ pennies

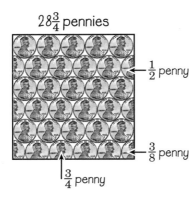

$\frac{1}{2}$ penny

$\frac{3}{8}$ penny

$\frac{3}{4}$ penny

Materials typing paper, optional (one ream per class); drawing paper, optional (10 sheets per group of students); transparency of a square with sides of 10 centimeters, optional (one per class); pennies, optional (one per student); meter sticks, optional (one per group of students)

Overview Students estimate whether one ream of typing paper will totally cover the floor of their classroom. They also estimate the number of pennies needed to cover a square with sides of 10 centimeters.

Planning Students may work in pairs or in small groups on problems **11–14.** You may want to use problem **12** for assessment. Problem **14** may be assigned as homework. Briefly discuss problem **12,** asking at least two students for their answers and reasoning. Problem **13** is optional.

Comments about the Problems

11. If students are having difficulty, ask them to find the thickness of 100 sheets of typing paper first and then use this answer to determine the thickness of one sheet.

12. Informal Assessment This problem assesses students' ability to use the concepts of perimeter, area, and volume to solve realistic problems.

Students may choose to cover the classroom floor with the typing paper or use meter sticks to measure the dimensions of the floor to determine its area and the quantity of typing paper needed to cover it.

14. Homework This problem may be assigned as homework. This activity implicitly shows how units other than squares may be used to measure area. This idea returns in problem **19** on page 36 of the Student Book.

You might have students mark off an area of paper that is 10 centimeters by 10 centimeters. Then have them model the situation using the two methods in which the pennies can be placed. (See the solutions column.) The method on the left seems to be the most obvious, but the other method leaves smaller empty spaces.

Population Density

The population density of the United States is estimated to be 70 people per square mile. This means that if all the people in the United States were spread out evenly over the whole country, there would be 70 people in each square mile. The population density of Egypt is estimated to be 154 people per square mile. The population density of The Netherlands is estimated to be 958 people per square mile.

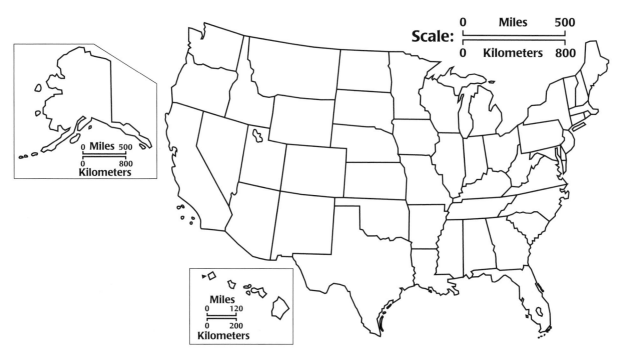

15. a. Estimate how many people could stand in your classroom at the same time.

   b. Compare your estimate with those of some of your classmates.

   c. If you think your answer to part **a** is a good estimate, write an explanation in your notebook or journal to defend it. Otherwise, rework your original estimate and describe your corrections.

16. a. If 10 people can stand in 1 square meter, how many people can stand in a square kilometer?

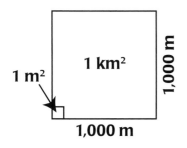

Drawing not to scale.

   b. Using 10 people per square meter, decide how big an area is needed for everyone in the world (about 5.6 billion people) to stand together.

   c. Would it be possible for all of the people in the world to stand in your state? Explain how you decided.

15. a. Answers will depend on the size of the classroom. Some students may mark a square meter on the floor and experiment to see how many people can stand in it. Other students may ask classmates to line up against the walls to find the length and width of the classroom in "classmate units."

b. Answers will vary. Sample student response:

I said 384 children could stand in our classroom, and one other classmate said 528. That's because when I asked children to stand in the square meter, only eight children fit. When my classmate did it, 11 children fit. The difference for one square meter was only three children, but the difference for 148 square meters was more than 100 children.

c. Answers will vary. Sample student responses:

I think I my estimate is good. Our classroom is 6 m × 8 m, or 48 m². I marked a square meter on the floor with tape and found out that 10 children can stand in this space. With the furniture removed, 480 children could stand in the classroom, but it would be very crowded.

My estimate is a good one. I used the children in my class as a measuring unit. The length of our classroom is 22 children and the width is 20 children, so 440 children can fit in the classroom if nothing else is in here.

16. a. 10,000,000 people

b. 560,000,000 square meters, which is equal to 560 square kilometers

c. Answers will vary. Sample student response:

Yes, all the people in the world can fit into our state. I looked in the almanac to find the area of our state and compared it to the answer to problem 16b.

Yes, the area of Georgia is about 4,500 square kilometers and that is much greater than 560 square kilometers. The approximate area is about $\frac{1}{2}$ × 300 km × 300 km, or 4,500 square kilometers.

Overview Students solve problems about population density.

About the Mathematics Population density is the ratio relationship between the number of people in a specific country and the area of that same country. This ratio is written as the number of people per square mile or per square kilometer. Ratio concepts are further developed in the grade 7/8 unit *Cereal Numbers*.

Planning Students may work in pairs or in small groups on problems **15** and **16**. Discuss problem **16** with the whole class. You can use problem **15** for assessment.

Comments about the Problems

15. Informal Assessment This problem assesses students' understanding of which units and tools are appropriate to estimate and measure area, perimeter, and volume and their ability to use the concepts of perimeter, area, and volume to solve realistic problems.

You may want to discuss the abbreviations for square meters (m²) and square kilometers (km²).

16. b. Some students may have difficulty writing 5.6 billion in standard form. You might suggest that students organize their computations in a ratio table like the one below.

Number of People	10	1,000,000,000	5,600,000,000
Area	1 m²	100,000,000 m²	560,000,000 m²

A Cloud of Snow Geese

Many birds migrate seasonally to avoid extreme weather. Above is a photo of a flock of snow geese.

17. a. How many geese are in this picture? Think of a way to estimate the number.

b. Compare your method with those of your classmates.

17. a. Estimates will vary. Accept estimates between 350 and 450 geese. Explanations will vary. Some students may use the number of snow geese in a small part of the picture to estimate the total number. To account for the variability in density, students may

- use a square grid and count the number of snow geese in an average square or a group of squares

- count a group of snow geese in an area of average density and use this area as a measuring unit.

b. Answers will vary. Sample student response:

Many of us said that about 400 snow geese were in the picture, but we estimated in different ways. I used an inch ruler and found the picture was $6\frac{3}{4}$ in. $\times$ $4\frac{1}{2}$ in. I picked a 1-inch square and counted 13 geese in this square. Then I multiplied $6\frac{3}{4} \times 4\frac{1}{2} \times 13 = 395$ geese. I rounded the number off to 400. My classmate traced the outline of the picture on a piece of paper, cut it out and folded it into eighths. Then he counted 53 geese in $\frac{1}{8}$ of the picture. Then he multiplied $53 \times \frac{1}{8} = 424$ geese, which he also rounded to 400.

Materials glass container filled with dry beans or jelly beans, optional (one per class)

Overview Students use different strategies to estimate the number of geese in a picture.

About the Mathematics Strategies for finding the number of birds are based on the concept of density. Calculating the number of birds in the flock based on the number of birds in one part of the flock assumes that the density of the part and the whole are the same. Because the part of the flock chosen influences the estimate of the whole flock, it is important to use a representative sample.

Planning You may define *migration* as "traveling periodically from one region or climate to another, usually for feeding or breeding purposes." You may want students to work individually on problem **17** and assign part **a** as homework. Have students discuss the strategies used.

Comments about the Problems

17. a. Homework Part **a** of this problem can be assigned as homework.

b. Emphasize the strategies used to solve this problem rather than students' answers. Comparing their estimation techniques and results may help students see alternative solution strategies.

Did You Know? Waterfowl take advantage of the abundant summer food supply in the northern tundra and migrate south to avoid the bitter winters. Some birds from arid countries migrate in order to find rain. The times of migration may be determined by hormones. The birds' internal clocks seem to be attuned to the lengths of the days and the climate.

Extension You might fill a glass container with dry beans or jelly beans and ask students to estimate the number of beans in the container. They can first estimate the number of beans lying on the bottom layer of the container. Then they can estimate the number of layers in the container.

Writing Opportunity Ask students to look for a picture of a large number of animals, people, or objects. They can write in their journals how they would estimate the total number.

Summary

Area is often expressed using standard units of measure.

Here are some metric measurements that describe area:
   square millimeters
   square centimeters
   square meters
   square kilometers

Here are some English measurements that describe area:
   square inches
   square feet
   square yards
   acres
   square miles

If a park has an area of 1 square kilometer, this does not mean the park is in the shape of a square. It may have any shape, as long as the area is equal to the area of a square measuring 1 kilometer by 1 kilometer.

You calculated population densities for people and geese. A *population density* is the number of members per unit of area.

Summary Questions

18. Name an object and estimate its area using at least two of the measurement units listed above.

19. Why do you think pennies do not work well as a unit of measurement for area?

18. Answers will vary. Sample student responses:

- A fingernail is about 1 square centimeter or about 100 square millimeters.

- A poster is about 1 square meter or about 9 square feet.

- A seat cushion is about 1 square foot or 144 square inches.

- Lake Tahoe is about 200 square miles or 700 square kilometers.

19. Circles are not a good unit for measuring area because no matter how they are arranged there are always spaces between them.

Overview Students read the Summary, which reviews the standard metric and English measurement units used to express area. Students select a shape and estimate its area using two different measurement units. They also explain why pennies do not work well as a measurement unit for area.

Planning Read and discuss the Summary as a class. Ask students to suggest other units of measurement. You may also want to discuss these common abbreviations:

- cm^2 = square centimeters
- m^2 = square meters
- km^2 = square kilometers

To determine whether or not students understand which units are appropriate for measuring various objects, ask the following questions:

- *What units are useful for measuring the length of a pencil?* [cm, mm, in.] Students' answers will show their understanding of the relationship between the size of the unit that is chosen and the precision of the measurement. You may point out that a millimeter is one-tenth of a centimeter.

- *Why is a meter or a kilometer not suitable for measuring a pencil?* [They are too large]

- *What unit is suitable for measuring the distance from home to school?* [kilometer or mile]

You may want students to work individually on problems **18** and **19.** Problem **18** can be used for assessment, and problem **19** can be assigned as homework. After students complete Section D, you may assign appropriate activities from the Try This! section, located on pages 49–52 of the Student Book, as homework.

Comments about the Problems

18. Informal Assessment This problem assesses students' understanding of which units and tools are appropriate to estimate and measure area, perimeter, and volume; and their understanding of the structure and use of standard systems of measurement, both metric and English.

19. Homework This problem may be assigned as homework. You may want to refer to problem **14** of this section. Stress that the area outside the pennies is waste and will not be reallotted.

Work Students Do

Students examine trails around lakes, garden fences, and enlargements of pictures to explore the relationship between perimeter and area. They then use the perimeter of a hexagon inscribed within a circle and that of a square circumscribed around a circle to help them informally discover the formula for the circumference of a circle and the value of π (≈ 3.14 or $\frac{22}{7}$). Students use cubes to build rectangular boxes of various shapes and fruit juice cans to explore the relationship between surface area and volume. Finally, students estimate the volume of a described snowfall to solve a problem about surface area and volume.

Goals

Students will:

• understand which units and tools are appropriate to estimate and measure area, perimeter, and volume;

• understand the structure and use of standard systems of measurement, both metric and English;

• use the concepts of perimeter, area, and volume to solve realistic problems;

• represent and solve problems using geometric models;

• analyze the effect a systematic change in dimension has on area, perimeter, and volume;

• generalize formulas and procedures for determining the areas of rectangles, triangles, parallelograms, quadrilaterals, and circles.

Pacing

• approximately four or five 45-minute class sessions

Vocabulary

• circumference
• diameter
• hexagon
• perimeter
• radius
• regular

About the Mathematics

This section focuses on the concepts of perimeter, area, and volume. The relationship between the perimeter and the area of a rectangle is explored: systematically doubling the perimeter of a rectangle increases its area by a factor of four. The circumference and the area of a circle are introduced using a diagram of a circle within which is a regular hexagon and around which is a square. The dimensions of the hexagon and the square are used to approximate the circle's circumference and to develop the formula $C = \pi \times d$. Enlargements of the hexagon-circle-square diagram are used to approximate the value of π (≈ 3.14 or $\frac{22}{7}$) as a number slightly greater than 3. The formula for the area of a circle, $A = \pi \times r \times r$ or $A = \pi r^2$, is developed by transforming a circle into a parallelogram.

The relationships between surface area and volume are introduced in this section. Rearranging a fixed number of cubes to build boxes of different shapes illustrates that a change in shape causes the surface area of the box to change. Fruit juice cans and their enlargements illustrate that this concept also applies to cylinders. Volume is expressed using both metric and English units.

Materials

- Student Activity Sheets 4 and 12, pages 128 and 136 of the Teacher Guide (one of each per student)
- string—25 and 50 cm long, pages 87 and 95 of the Teacher Guide, optional (one piece of each per group of students)
- grid paper, pages 87 and 107 of the Teacher Guide, optional (one or two sheets per student)
- transparency of grid paper, page 87 of the Teacher Guide, optional (one sheet per class)
- overhead projector, pages 87, 89, 91, and 97 of the Teacher Guide, optional (one per class)
- transparency of the two rectangles on page 91 of the Teacher Guide, optional (one per class)
- centimeter rulers, pages 89, 91, 93, and 95 of the Teacher Guide (one per student)
- circular objects, page 95 of the Teacher Guide (three objects per group of students)
- compasses, pages 95 and 101 of the Teacher Guide (one per student)
- protractors, page 95 of the Teacher Guide, optional (one per student)
- metric measuring tapes, page 95 of the Teacher Guide, optional (one per group of students)
- scissors, page 97, 101, and 103 of the Teacher Guide (one pair per student)
- centimeter cubes, page 99 of the Teacher Guide (24 per pair or group of students)
- toilet paper rolls, page 101 of the Teacher Guide (two per group of students)
- cans, page 101 of the Teacher Guide (two per group of students)
- scrap paper, page 103 of the Teacher Guide (two sheets per group of students)
- tracing paper, page 107 of the Teacher Guide, optional (one sheet per group of students)

Planning Instruction

In this section, students extend their understanding of the relationships among perimeter, area, and volume. It is important for students to discuss these mathematical relationships as a whole class. Some students may have difficulty using numerical formulas for area and volume. Allow these students to use word formulas or estimations.

Problems 1–2, 17, and 27 are whole-class activities. Students may work individually on problems 7, 12, 19, and 29–31. The remaining problems can be done in pairs or in small groups.

Problems 20 and 26 are optional, and you may consider problem 28 optional if it is too difficult for your students. If time is a concern, you may omit these problems or assign them as homework.

Homework

Problems 7 (page 90 of the Teacher Guide) and 26 (page 102 of the Teacher Guide) can be assigned as homework. The Extensions (pages 87, 89, and 107 of the Teacher Guide) can also be assigned as homework. After students complete Section E, you may assign appropriate activities from the Try This! section, located on pages 49–52 of the Student Book. The Try This! activities reinforce the key mathematical concepts introduced in this section.

Planning Assessment

- Problems 6 and 31 can be used to informally assess students' ability to analyze the effect a systematic change in dimension has on perimeter, area, and volume.
- Problem 19 can be used to informally assess students' ability to generalize formulas and procedures for determining the areas of rectangles, triangles, parallelograms, quadrilaterals, and circles.
- Problems 24, 27b, 29, and 30 can be used to informally assess students' understanding of the structure and use of standard systems of measurement, both metric and English.
- Problem 28 can be used to informally assess students' ability to use concepts of perimeter, area, and volume to solve realistic problems and to represent and solve problems using geometric models.
- Problems 29 and 30 can be used to informally assess students' ability to understand which units and tools are appropriate to estimate and compute perimeter, area, and volume.

E. PERIMETER, AREA, AND VOLUME

Perimeter

Danny is a contractor who has been hired to build a bicycle/running trail around each of the lakes pictured below. The owner wants to pay Danny the same amount of money for each trail because the lakes are equal in area. Danny agrees that the lakes are equal in area, but he wants more money for constructing the trail around Lake Marie.

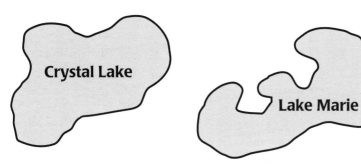

1. Do you agree that Danny should get more money for the Lake Marie trail? Why or why not?

Below are diagrams of four gardens the city wants to plant in a downtown park. Along the outside edges of each garden, the city will place an ornamental fence.

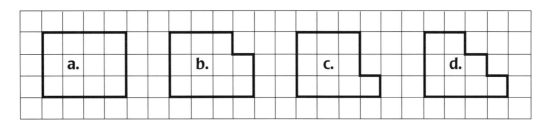

2. How much fencing is needed for each garden?

The distance around a shape is called the *perimeter* of the shape.

3. Find the area of each garden.

4. Describe any patterns that you can find in the areas and perimeters of these four gardens.

Below is a diagram of a garden with an area of 15 square units.

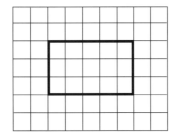

5. Use **Student Activity Sheet 4** or grid paper to design four other gardens with the same area but with different perimeters. Label the area and perimeter of each garden.

1. Yes, Danny should get more money for the Lake Marie trail because Lake Marie has a larger perimeter.

2. Each garden will need 14 units of fencing.

3. **a.** 12 square units

   b. 11 square units

   c. 10 square units

   d. 9 square units

4. Answers will vary. Sample student response:

   The perimeter of each garden is 14 units, but the areas decrease by one square unit as you move from left to right.

5. Drawings will vary. Sample student response:

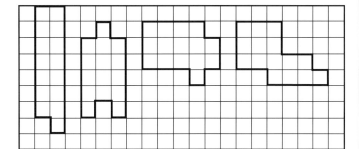

Materials Student Activity Sheet 4 (one per student); string—25 cm long, optional (one per student); grid paper, optional (one or two sheets per student); transparency of grid paper, optional (one sheet per class); overhead projector, optional (one per class)

Overview Students compare the length of a trail around one lake to the length of a trail around another lake. They then measure and compare the perimeters and areas of different-shaped gardens drawn on grid paper.

About the Mathematics The contexts of trails and fences help students focus on the concept of perimeter. In problems **4** and **5,** students discover that figures with identical perimeters can have different areas and that figures with identical areas can have different perimeters.

Planning Complete problems **1** and **2** as a whole class. After problem **2,** ask students to explain the terms *perimeter* and *area.* Encourage them to use their own words and illustrate their explanations with examples. Students can work on problems **3–5** in pairs or in small groups.

Comments about the Problems

1. Students may use a piece of string to verify their answers.

2–3. Suggest that students use one side of a grid square as a measuring unit. Some students will incorrectly include a specific unit of measure in their answer. Point out that the grids have no specific unit of measure.

Extension Have students use graph paper to draw different-shaped gardens with areas of 15 units. Then ask them to find the perimeter of each garden. Challenge students to find the rectangular and square gardens that have the largest possible perimeters. You may want to draw these gardens on a transparency of grid paper. Two possibilities are illustrated below.

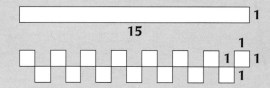

Below are three pairs of figures. The figures in each pair are identical in shape but not in size.

For each pair, figure **i** has been enlarged to make figure **ii.**

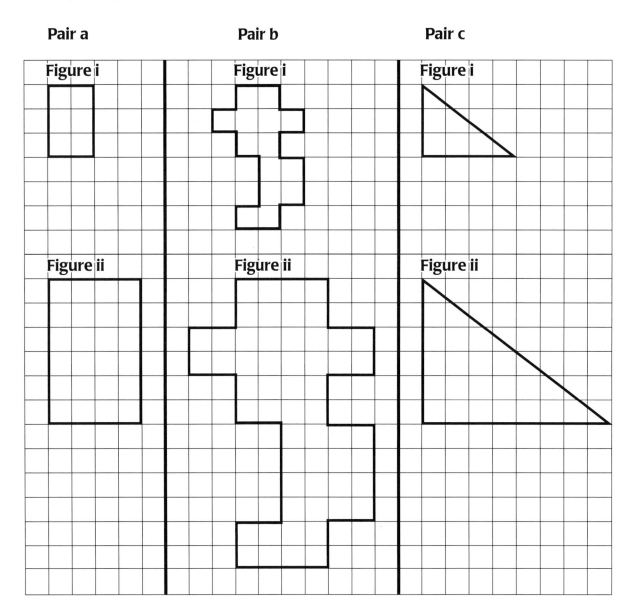

Pair a **Pair b** **Pair c**

Figure i Figure i Figure i

Figure ii Figure ii Figure ii

6. Investigate the effect that this enlargement has on area and perimeter. Compare the areas and perimeters of figure **i** and figure **ii** in each pair shown above.

6. In each pair of figures, each side of figure i has been doubled to get figure ii. The area of figure ii is four times that of figure i. The perimeter of figure ii is twice that of figure i.

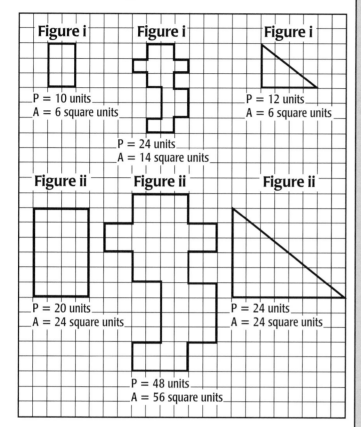

Figure i
P = 10 units
A = 6 square units

Figure i
P = 24 units
A = 14 square units

Figure i
P = 12 units
A = 6 square units

Figure ii
P = 20 units
A = 24 square units

Figure ii
P = 48 units
A = 56 square units

Figure ii
P = 24 units
A = 24 square units

Materials centimeter rulers, optional (one per student); transparency of the two rectangles on this page, optional (one per class); overhead projector, optional (one per class)

Overview Students compare the perimeter and area of a figure with those of its enlargement.

About the Mathematics This activity helps students understand that area and perimeter are not directly proportional. In the grade 5/6 unit *Grasping Sizes*, students learned that the ratios between measurements of a shape and those of its enlargement are constant. For example, consider the ratio relationships between these figures.

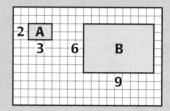

$$\frac{\text{height A}}{\text{height B}} = \frac{\text{base A}}{\text{base B}} = \frac{1}{3}$$

Also consider the ratio relationship between two sides of the first figure and the two corresponding sides of the second figure.

$$\frac{\text{height A}}{\text{base A}} = \frac{\text{height B}}{\text{base B}} = \frac{2}{3}$$

If the corresponding sides of a figure and its enlargement do not have the same ratio, the original figure has been distorted.

Planning Students can work in pairs or in small groups on this problem. Be sure to discuss the responses in class. You may use problem **6** for assessment.

Comments about the Problems

6. Informal Assessment This problem assesses students' ability to analyze the effect a systematic change in dimensions has on perimeter, area, and volume.

Extension Ask students to choose the smaller figure from one of the three pairs on Student Book page 38 and enlarge it by making the sides three times longer. Then ask students to investigate the effect of this enlargement on the figure's area and perimeter.

d.

c.

b.

a.

Area

Different enlargements of the same print have the same shape.

The best way to show this is to place the pictures on top of each other, with the lower, left-hand corners at the same spot.

The top, right-hand corners can be connected with a straight line.

You can use this line to construct a new rectangle that has the same shape.

This method can also be used to outline a portion of the print that can be enlarged to make a new print of the same shape.

d.

c.

b.

a.

7. Measure the base and height in centimeters of each of the photos **(a–d)** shown above.

7. Students' answers may vary, depending upon their ability to make precise measurements. Accept answers within two millimeters of the given answers in the table below.

Picture	Base (in cm)	Height (in cm)
a	2.2	1.5
b	4.5	3.0
c	6.8	4.6
d	9.0	6.2

Materials centimeter rulers (one per student); overhead projector, optional (one per class); transparency of the two rectangles on this page, optional (one per class)

Overview Students investigate different enlargements of the same picture and measure the base and height of each.

Planning You may assign problem **7** for homework and discuss it the next day in class. Focus on the relationships between the measurements and on the fact that the pictures have the same shape. This discussion will prepare students for problems **9–11** on the next page.

Comments about the Problems

7. Homework This problem may be assigned as homework. Encourage students to record their measurements in a chart so they can more easily see the relationships between the measurements of corresponding sides.

Students may notice several relationships. For example, the base and the height of print **d** are about two times longer than the base and height of print **b;** the base of each print is about one and a half times longer than its height. Answers may vary depending on students' precision in measuring and on whether they round off their measurements.

Almost all students will notice that a shape changes when a rectangle's height is enlarged by more than its width. However, some students may say that the shape remains the same because it is still a rectangle. To clear up this misconception, you might use the chalkboard or make a transparency of the rectangles below. Ask students whether Figure A could be enlarged to produce Figure B.

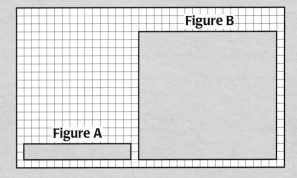

Here is an original print.

8. What are the measurements of the print?

9. Below is a table of measurements for some enlargements. The heights have been given to you. What are the base measurements?

Height	Base
12 cm	?
18 cm	?
24 cm	?

10. a. A piece of glass for a print that measures 12 centimeters by 18 centimeters costs $5. What will the price of the glass be if the picture is enlarged to 24 centimeters by 36 centimeters?

b. Wood molding to make a frame for this 12-by-18-centimeter print costs $10. What will the molding cost for the 24-by-36-centimeter enlargement?

11. The print below has been enlarged from the original above by a factor of two. This means that both the length and width have doubled. How has the area changed?

Solutions and Samples
of student work

8. height = 6 cm
base = 8 cm

9.

Height (in cm)	Base (in cm)
12	16
18	24
24	32

10. a. $20. If the height and base are doubled, the area will increase by a factor of four.

b. $20. If the height and base are doubled, the perimeter will increase by a factor of two.

11. The area increased by a factor of four.

Hints and Comments

Materials centimeter rulers (one per student)

Overview Students investigate three different enlargements of the same print. Using a given height, they find the base measurement of each enlargement. They then determine how the area of a print changes when the print is enlarged by a factor of two.

About the Mathematics In problem **9,** the prints are enlargements of one original print, so the table is a ratio table. To find unknown measurements in a ratio table, you can use the relationship between the rows or the columns. Students can compare the dimensions within a single rectangle to find a ratio relationship. They may also compare the corresponding dimensions of two different rectangles to find other ratios. In the following table, a comparison across each row shows that the ratio between the base and height within each rectangle is 6 to 9, or 2 to 3.

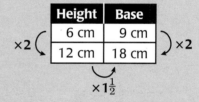

Planning Students can work on problems **8–11** in pairs or in small groups. Discuss how to find an unknown measurement in a ratio table (as in problem **9**) and the effect of enlargement on area (as in problem **11**).

Comments about the Problems

9. If students have difficulty, suggest that they include the measurements of the original print in the ratio table so they can more easily see the ratio relationships.

10. The price of the glass depends on the area of the print, and the price of the frame depends on the perimeter of the print. Some students may draw the enlarged print to help them see the difference in the bases of the price for the glass and the price for the frame.

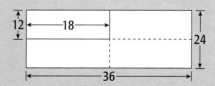

On the right is a regular triangle. *Regular* means that all sides have the same length.

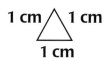 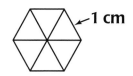

Using six of these regular triangles, you can make a regular hexagon. A *hexagon* is a six-sided figure.

12. a. What is the perimeter of this regular hexagon?

   b. How can you draw a circle that passes through all corners of the hexagon?

   c. What are the dimensions of the smallest square that would totally enclose the hexagon and circle?

   d. In your notebook, draw the hexagon and circle enclosed in a square.

13. Use a compass to create three enlargements of the shapes you drew for problem **12d.** Use regular triangles with 2-centimeter sides for one hexagon. Choose two larger lengths for the other two hexagons. Then enclose each hexagon with a circle and a square.

14. Copy the table below into your notebook. Use the figures you constructed to fill in the table. The perimeters of the squares and hexagons can be found without measuring.

The *diameter* of a circle is the length of a straight line through the center of the circle.

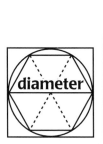

	Lengths			
	Using 1-cm Triangles	Using 2-cm Triangles	Using __-cm Triangles	Using __-cm Triangles
Diameter of Circle				
Perimeter of Hexagon				
Approximate Perimeter of Circle				
Perimeter of Square				

15. Look at the perimeters. How are the perimeters of the circles related to the perimeters of the hexagons and the squares?

The perimeter of a circle has a special name—the *circumference.*

16. a. In the classroom, find three different circular objects and measure their diameters.

   b. Estimate the circumferences of these objects. Check your estimates by measuring the circumferences using string or a tape measure.

12. a. 6 centimeters

b. Answers will vary. Sample student response:

Put the point of a compass at the center of the hexagon and set the radius equal to the length of a side of the hexagon.

c. 2 cm × 2 cm

d. The drawing should look like those in the solution to problem **13** and have a radius of 1 cm.

13.

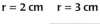

r = 2 cm   r = 3 cm   r = 4 cm

14. One possible table follows:

	Lengths			
	Using 1-cm Triangles	Using 2-cm Triangles	Using 3-cm Triangles	Using 4-cm Triangles
Diameter of Circle	2 cm	4 cm	6 cm	8 cm
Perimeter of Hexagon	6 cm	12 cm	18 cm	24 cm
Approximate Perimeter of Circle	a little more than 6 cm	a little more than 12 cm	a little more than 18 cm	a little more than 24 cm
Perimeter of Square	8 cm	16 cm	24 cm	32 cm

15. Answers will vary. Sample student response:

The perimeter of the circle is a little bit larger than the perimeter of the hexagon, but it's smaller than the perimeter of the square. The perimeter of the circle is a little more than $\frac{3}{4}$ the perimeter of the square.

16. a. Answers will vary. Sample student responses:

cup: 8 centimeters
flower pot: 15 centimeters
trash can: 40 centimeters

b. Estimates will vary. Sample student responses:

Object	Diameter	Estimated Circumference
cup	8 cm	25 cm
flower pot	15 cm	47 cm
trash can	40 cm	130 cm

Materials circular objects (three per group); string–50 cm long, optional (one per group); compasses (one per student); protractors, optional (one per student); metric measuring tapes, optional (one per group); centimeter rulers (one per student)

Overview Students draw regular hexagons of different sizes and enclose each hexagon within a circle and a square. They then investigate the perimeters of these shapes.

About the Mathematics The drawing below shows that the estimated circumference of the circle is between the perimeter of the hexagon and the perimeter of the square. The perimeters of the shapes can be expressed in a more formal way using the diameter (*d*) of the circle: the perimeter of the square is 4 × *d* and the perimeter of the hexagon is 3 × *d*. The picture shows that the circumference of the circle is closer to 3 × *d* than 4 × *d*.

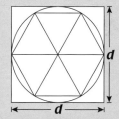

Planning You might want to discuss the terms *diameter*, *radius*, and *hexagon* before students begin problem **12**. Students can work in pairs or in small groups on problems **13–16**.

Comments about the Problems

12. d. Some students may have difficulty making this drawing. Some students may recall that each angle in a regular triangle is 60° and use a compass and a protractor to draw these figures. Other students may use the diagram of the 1-centimeter triangle and hexagon on page 41 of the Student Book as a template.

14. Students should estimate that the perimeter of the circle is less than the perimeter of the square and greater than the perimeter of the hexagon.

16. b. Some students may see that the circle's circumference is a little more than three times its diameter. If students do not see this relationship, refer them to the table in problem **14**.

Long ago, people discovered that the ratio of the circumference of a circle to the diameter of that circle is always the same. The Greek letter "π" (pronounced pī) is used to represent this ratio. It is approximately 3.14, or $\frac{22}{7}$.

A final column can be added to the table in problem **14.** This column gives the general formulas you may have discovered in your work on page 41.

Descriptions	Unknown Lengths
Diameter of Circle	d
Perimeter of Hexagon	$3 \times d$
Circumference or Perimeter of Circle	$3.14 \times d$ or $\pi \times d$
Perimeter of Square	$4 \times d$

The circumference of a circle can also be written in terms of the radius of the circle. The *radius* is one-half of the diameter.

17. **a.** Write a formula relating the radius and diameter.

   b. Write a formula for the circumference of a circle in terms of the radius.

18. Cut apart the sections of the circle on **Student Activity Sheet 12.** Rearrange them to help you develop a formula for the area of a circle.

19. Write a paragraph describing what you now know about the circumference and area of any circle.

20. There are many different uses for glass. Below are some designs using glass, along with prices per square foot. For each picture, calculate the total cost. If a design includes trim, the total cost should include this.

TABLE TOP
Sheet Glass:
$4.00 per
square foot

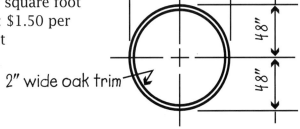

CIRCLEHEAD WINDOW
Window Glass:
$3.50 per square foot
Maple Trim: $1.50 per
linear foot

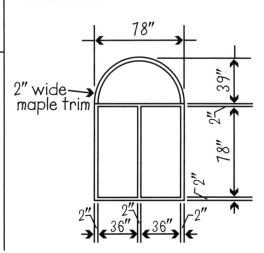

CIRCULAR WINDOW
Window Glass:
$3.50 per square foot
Oak Trim: $1.50 per
linear foot

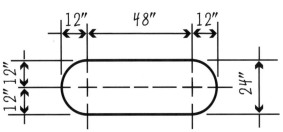

2" wide oak trim

17. a. $d = 2 \times r$, or $r = \frac{1}{2} \times d$

b. Answers will vary. Some possibilities are
$C = 2 \times r \times \pi$ or $(r + r)\,\pi$ or $2 \times 3.14 \times r$

18. See the drawing for problem **18** in the Hints and Comments column on this page. Students' approaches will vary. The area of a circle is approximately the area of a parallelogram with a height about equal to its radius and a base about equal to one-half its circumference, or
$A = r\left(\frac{1}{2} \times 2 \times \pi \times r\right) = \pi \times r \times r = \pi r^2$

19. Answers will vary. Sample student response:

The distance around the outside of the circle is the circumference. The formula is $C = \pi d$.

To find the area of a circle, I need to know the length of the radius, which is one-half the length of the diameter. I use the formula $A = \pi r^2$ for area.

20. TABLETOP: Total Cost = $44.56
Area of Rectangle = 2 ft × 4 ft = 8 ft^2
Cost = 8 × $4.00 = $32.00

Area of 2 semicircles = 1 circle in which $r = 1$ ft
$A = \pi r^2 = \pi \times 1 \text{ ft} \times 1 \text{ ft} = 3.14 \text{ ft}^2$

Cost = 3.14 × $4.00 = $12.56

CIRCULAR WINDOW: Total Cost = $213.54
Glass: Area of circle in which $r = 4$ feet
$A = \pi r^2 = \pi \times 4 \times 4 = 50.24 \text{ ft}^2$
Cost of glass = 50.27 × $3.50 = $175.84

Trim: Perimeter of circle in which $d = 8$ ft
Circumference = $\pi \times d$ = 25.12 ft

Cost of trim = 25.13 × $1.50 = $37.70

CIRCLEHEAD WINDOW: Total cost = $252.05
Glass: 2 Rectangles, each 3 ft × 6.5 ft
Area = 2 × 3 ft × 6.5 ft = 39 ft^2
Cost of glass = 39 ft^2 × $3.50 = $136.50

Semicircle (half of 1 circle), $r = 37$ in.
$A = \frac{1}{2} \times \pi r^2 = \frac{1}{2}\pi \times (37 \text{ in.})^2 = 2{,}149.33 \text{ in.}^2$
2,149.33 in.^{2} ÷ 144 = 14.93 ft^2
Cost of glass = 14.93 × $3.50 = $52.26

Trim: Square, 1 side = 78 in. and Center strip = 78 in.
$P = 4 \times 78$ in. = 312 in.
78 in. + 312 in. = 390 in. = 32.5 ft

Semicircle in which $d = 74$ in.
$C = \frac{1}{2} \times \pi d = \frac{1}{2} \times \pi \times 74$ in. = 116.18 in. = 9.69 ft

Trim needed = 32.5 ft + 9.69 ft = 42.19 ft
Cost of trim = $1.50 per ft × 42.19 ft = $63.29

Materials Student Activity Sheet 12 (one per student); scissors (one pair per student); overhead projector, optional (one per class)

Overview Using the relationship between the radius and circumference of a circle, students discover and apply formulas for finding the area and circumference of a circle.

About the Mathematics In a circle, the ratio of the circumference to the diameter is constant. Students can discover this ratio by dividing a circle's circumference by its diameter to get a number a little greater than 3. This ratio is commonly referred to as π ($\pi = \frac{C}{d}$).

Planning Ask students to explain their estimates for problem **16b** on the previous page. When they say *something between 3 and 4 times the diameter,* you can ask them, *Will it be between 3 and 3.5 times or between 3.5 and 4 times?* [between 3 and 3.5 times] Then ask students to explain each formula in the table on page 42 of the Student Book. Students who have had little experience with formulas may need to use word formulas to express the relationships.

Complete problem **17** as a whole class. Students can work on problem **18** in pairs or in small groups. Discuss problem **18** before continuing with problem **19,** which can also be used for informal assessment. Problem **20** is optional.

Comments about the Problems

17. b. Students may answer with word formulas.

18. Suggest that students transform the circle into a parallelogram. You may demonstrate this transformation by cutting apart the segments of the circle on Student Activity Sheet 12 and rearranging them on an overhead projector to form a parallelogram.

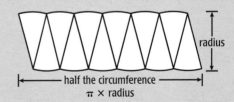

half the circumference
π × radius

19. Informal Assessment This problem assesses students' ability to generalize formulas and procedures for determining areas of rectangles, triangles, parallelograms, quadrilaterals, and circles. Allow students to demonstrate their strategy with an example.

Activity

Volume and Surface Area

You have investigated the relationship between perimeter and area. In a similar way, you can look at the relationship between volume and surface area. This can be thought of as the relationship between the space inside a package and the area that can be covered by its wrapping.

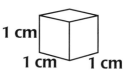

1 cm
1 cm   1 cm

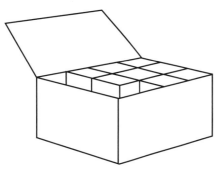

A rectangular box that will hold exactly 24 cubes is said to have a volume of 24 cubic centimeters, abbreviated as 24 cm^3.

21. Use centimeter cubes to find as many different-sized boxes as you can that will hold exactly 24 cubes. The dimensions of each box should be whole numbers. Find out how much cardboard would be needed to make each box (this is the surface area). Use a table like the one below to record your results.

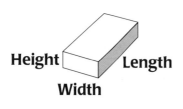

Height   Length
Width

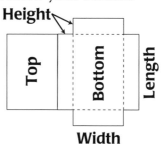

Height
Top   Bottom   Length
Width

Length (in cm)	Width (in cm)	Height (in cm)	Volume (in cm^3)	Surface Area (in cm^2)
			24	
			24	
			24	
			24	
			24	
			24	
			24	

21. Tables will vary. Sample student response:

Length (in cm)	Width (in cm)	Height (in cm)	Volume (in cm³)	Surface Area (in cm²)
12	2	1	24	76
6	4	1	24	68
4	3	2	24	52
6	2	2	24	56
24	1	1	24	98
8	3	1	24	70

Materials centimeter cubes, optional (24 per pair or group of students)

Overview Students create different rectangular boxes, all with the same volume, and find the surface area of each.

About the Mathematics Area can be used to describe the surface area of a three-dimensional solid as well as a two-dimensional shape. This activity illustrates that volume and surface area are not directly proportional. For example, the two solids below each have a volume of 8 cubic units. Solid A has a surface area of 34 square units, while the surface area of Solid B is 24 square units.

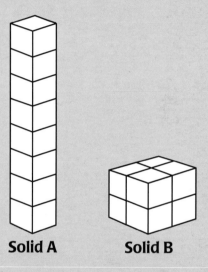

Solid A   **Solid B**

Planning Have students work in pairs or small groups on problem **21.** Discuss students' results in class.

Comments about the Problems

21. Encourage students to use the centimeter cubes. Making a table to systematically list the solutions can help students find all the possible different-sized boxes.

Ask students to think about the box with the smallest surface area and the one with the largest surface area. (The more a box resembles a cube, the smaller its surface area.)

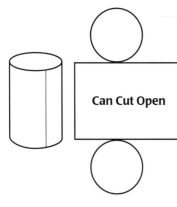

Can Cut Open

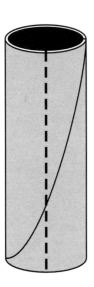

Fruit drinks come in cans of different sizes. Some cans are narrow and tall; others are wide and short. A juice can is made up of two circles and a rectangle. Cans that look completely different may contain the same amount of liquid or hold different amounts.

22. The type of fruit drink pictured on the left is also available in cans that are twice as high as the can you see in the picture.

    a. How do the amounts of liquid that fit in the two cans compare?

    b. What do you know about how the surface areas of the cans compare? Be prepared to explain your answer without making calculations.

23. Suppose one can has double the diameter of another can.

    a. Do you think the amount of liquid that fits in the larger can will be double? Be prepared to explain your answer.

    b. What can you tell about the surface area of the larger can compared to that of the original can?

Metro's Grocery sells juice in liter bottles. Many producers are switching to the metric system for measuring and bottling liquids. Any 10-centimeter cube will hold exactly 1 liter of liquid.

24. a. How many cubic centimeters are in one liter?

    b. Describe the volume of 3 liters of juice in two different ways.

Collect cardboard tubes from inside several rolls of toilet paper.

25. a. What kind of shape do you get when you cut a tube along its spiral line? Make a drawing.

    b. Cut another tube along a straight line, as shown in the picture on the left. Make a drawing of the resulting shape.

The two shapes you cut the tube into show that the area of a parallelogram is equal to the area of a rectangle with the same base and height. The shape of the parallelogram that is created by cutting the cylinder depends on the steepness of the cutting line. Still, no matter how steep you make the cutting line, the base, the height, and the total surface area will be the same.

22. a. Answers will vary. Sample student response:

The taller can holds twice as much juice as the shorter can.

b. Answers will vary. Sample student response:

The surface area of the taller can is a little less than two times the surface area of the smaller can. The rectangular part of the surface is twice as big, but the ends of the can are the same size.

23. a. Answers will vary. Sample student response:

No. If you double the diameter, the can will hold more than twice as much juice. This drawing shows why:

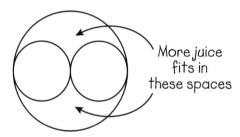

More juice fits in these spaces

b. Answers will vary. Sample student response:

Doubling the diameter makes the surface area much larger. The area of the side doubles, but the areas of the top and bottom are four times larger.

24. a. 1,000 cm³

b. Answers will vary. Sample student response:

Three liters of juice is about $\frac{3}{4}$ gallon. This amount of juice would fill a box measuring 10 centimeters by 10 centimeters by 30 centimeters.

25. a. a parallelogram

b. a rectangle

Materials toilet paper rolls (two per group of students); scissors (one pair per group of students); cans (two per group of students); compasses, optional (one per pair or group of students)

Overview Students investigate the volumes and surface areas of cylinders using cans and toilet paper rolls. They explore the effect of doubling a cylinder's height and then doubling its diameter.

Planning Have students work in pairs or in small groups on problems **22–25.** Then discuss students' answers in class. You may want to use problem **24** for assessment.

Comments about the Problems

22–23. Students are not expected to accurately compute the surface area and volume. It is more important for them to reason as they explore relative sizes.

22. It may be helpful for students to physically place one can on top of another to see that the volume doubles.

23. If students have difficulty, suggest that they draw circles to help them visualize the two cylinders. They can trace the end of the can to create the smaller circle. Then they can use a compass to draw the larger circle.

24. Informal Assessment This problem assesses students' ability to understand the structure and use of standard systems of measurement, both metric and English.

25. Let students cut apart the cardboard tubes to transform them into parallelograms and rectangles. Students revisit strategies for determining the area of a parallelogram.

Paper Patterns

Here you see paper patterns for a jacket and a skirt. The patterns are placed on a piece of folded fabric. Then the fabric is cut out using the patterns as a guide.

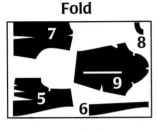

Fold

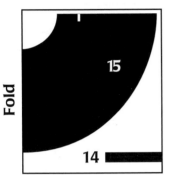

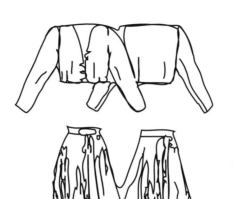

This picture shows that a sleeve looks like a cylinder with a top that angles, and the skirt looks like a cone without a top.

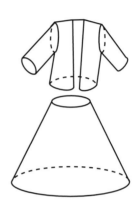

26. a. Make a paper cylinder and a paper cone to investigate the shapes of paper patterns for a sleeve and a skirt.

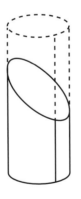

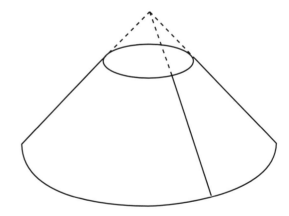

b. Now look at the paper patterns and indicate the numbers of the following parts:

waistband ___?___   sleeve ___?___

shirt front ___?___   shirt back ___?___   skirt ___?___

c. As you can see, a lot of cloth is wasted. Estimate what fraction of the cloth is used.

26. a. Patterns will vary. Sample student response:

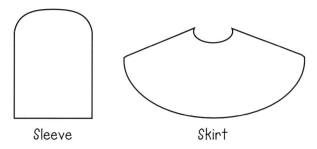

Sleeve   Skirt

b. Waistband, part 14
Sleeve, part 9
Shirt Front, part 5
Shirt Back, part 7
Skirt, part 15

c. Estimates will vary. Accept estimates between
$\frac{1}{3}$ and $\frac{1}{2}$.

Materials scrap paper (two sheets per group of students); scissors (one pair per student)

Overview Students solve problems involving sewing patterns.

About the Mathematics Sewing patterns show how flat shapes are used to construct three-dimensional shapes. From the grade 5/6 unit *Figuring All the Angles,* students may remember making a flat map to represent the three-dimensional Earth. Constructing nets of different geometric three-dimensional shapes is investigated in the grade 7/8 unit *Packages and Polygons.*

Planning Problem **26** is optional. If time is a concern, you may omit this problem or assign it as homework.

Comments about the Problems

26. a. Homework This problem may be assigned as homework. Students may find it interesting to make the size of the paper cylinder large enough so that they can wear it.

b. Homework This problem may be assigned as homework. You may ask your students what pieces numbers 6 and 8 are [Pieces 6 and 8 form the facing, which goes on the underside of the front and back of the jacket].

Snowfall

On January 27, 1967, Chicago had a terrible snowstorm that lasted 29 hours. Although January in Chicago is usually cold and snowy, 60 centimeters of snow from one snowstorm is unusual. For three days, the buses stopped. No trains ran. There was no garbage collection or mail delivery. Few people went to work. Most stores were closed. Chicago looked like a ghost town.

27. a. If the snowstorm lasted 29 hours, why was the city affected for three days?

    b. How much is 60 centimeters of snow?

    c. Has the town where you live ever had really bad weather? Describe what happened and what effect the weather had on your town.

27. a. Answers will vary. Sample student response:

Cars and buses can't drive through snow 60 centimeters deep. People had to wait until the snow was removed from the streets.

b. Answers will vary. Sample student responses:

- 60 cm of snow is about the length of my leg.

- 60 cm of snow is about the diameter of a car tire.

c. Answers will depend on students' experiences. Sample student response:

I live in Friars Point, Mississippi. It is on the Mississippi River. In April of 1993, it began to rain. It kept raining, and in early June, the river flooded. We tried to use sandbags to stop the flooding, but there was too much water. The rain continued through June, and many farms and houses were destroyed.

Overview On this page, students are introduced to the context of the final problem of this section: snowfall.

Planning You can ask students to look in a newspaper for weather reports, and to bring them to class. Rainfall and snowfall articles connect well with the problem on the next page. Complete problem **27** as a whole class. You may use problem **27b** for assessment. Briefly discuss students' answers in class.

Comments about the Problems

27. b. Informal Assessment This problem assesses students' ability to understand the structure and use of the standard metric and English systems of measurement.

As a whole class, discuss whether the accumulation of snow was 60 centimeters throughout the city, and how people determined this accumulation. In this text, the measurements of precipitation are given in centimeters. You may want to compare this unit of measurement to those in the newspaper articles students bring to class.

28. Use the map below to estimate the volume of snow that buried Chicago on January 27, 1967.

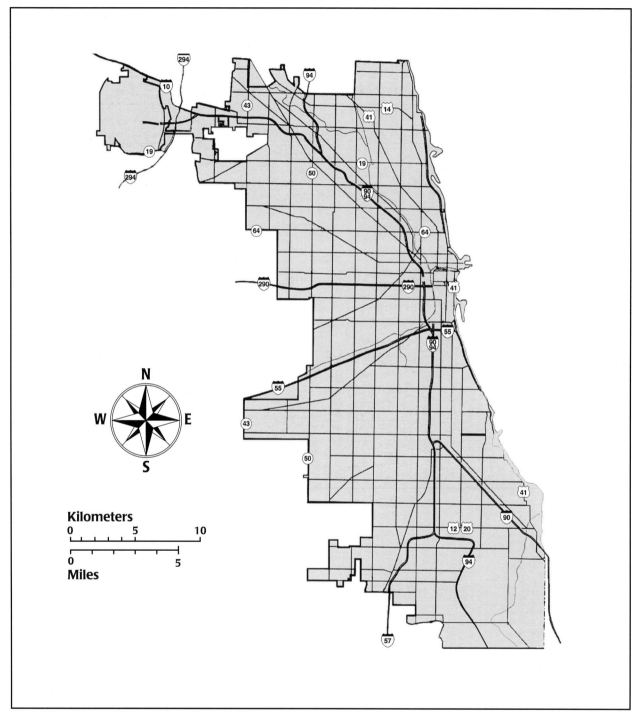

Courtesy of DonTech, publishers of the Ameritech PagesPlus® Yellow Pages.

Solutions and Samples
of student work

Hints and Comments

28. Estimates will depend on the estimated area of Chicago. Accept student estimates for the area of Chicago from 600 square kilometers to 800 square kilometers, Accept estimates for the volume of snow from 360,000,000,000,000 cubic centimeters to 480,000,000,000,000 cubic centimeters.

Sample student reasoning:

The area of Chicago is about 20 km × 40 km. The snow volume is 60 cm × 20 km × 40 km. Change all the dimensions to centimeters (1 km = 100,000 cm) to get 60 cm × 2,000,000 cm × 4,000,000 cm, which is 48,000,000,000,000,000 cm³.

Materials grid paper, optional (one sheet per group of students); tracing paper, optional (one sheet per group of students)

Overview Students estimate the volume of snow that fell in Chicago on a specific date.

About the Mathematics The strategy of *reallotment* may be used to calculate the volume in much the same way as it was previously used to calculate area. Irregularly shaped solids may be reallotted to form a rectangular box, which students can measure with a convenient unit of volume.

Planning Problem **28** can be used for informal assessment. If you think problem **28** is too difficult for your students, consider it optional. The Extension can be done as a whole-class activity.

Comments about the Problems

28. Informal Assessment This problem assesses students' ability to represent and solve problems using geometric models. It also assesses students' ability to use the concepts of perimeter, area, and volume to solve realistic problems.

Students can estimate the area of Chicago using grid paper, tracing paper, and any one of the strategies developed earlier in the unit.

Extension Present students with the following problem: *Suppose the area of Chicago were two times larger and the same amount (volume) of snow was spread out over that larger area. What would the accumulation of snow be then?* [The accumulation would be 30 centimeters, half of 60 centimeters.]

Summary

In this section, you studied perimeter and area and how they are related.

You saw that
• two shapes with the same perimeter can have different areas, and
• two shapes with the same area can have different perimeters.

You also studied volume and surface area and the relationship between them. You learned that two objects with the same volume can have different surface areas.

You found the formula for the area of a circle:

$$A = \pi \times r^2$$

where A represents the area, r represents the length of the radius, and π is approximately 3.14 or $\frac{22}{7}$.

The formula for the circumference (perimeter) of a circle is

$$C = \pi \times d$$

where C represents the circumference, and d represents the length of the diameter.

Summary Questions

29. Draw at least three different rectangles that have an area of 16 square centimeters. What is the perimeter of each rectangle?

30. A box can hold a maximum of 18 one-centimeter cubes. What are three possible surface areas for this box? Justify your answers.

31. a. What happens to the circumference of a circle if you multiply the diameter by two? Justify your answer.

b. What happens to the area of a circle if you multiply the diameter by two? Justify your answer.

29. Drawings will vary, and perimeters will vary based on the drawings. Sample student response:

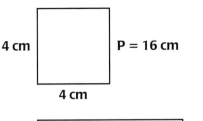

4 cm   P = 16 cm
4 cm

2 cm   P = 20 cm
8 cm

P = 34 cm
1 cm
16 cm

30. Answers will vary. (*Note:* Some students may give dimensions of irregularly shaped boxes.) Sample student responses:

Dimensions	Surface Area
6 cm × 3 cm × 1 cm	54 cm²
2 cm × 3 cm × 3 cm	42 cm²
9 cm × 2 cm × 1 cm	58 cm²

31. a. The circumference is doubled. Explanations will vary. Sample student response:

> If you double the diameter, the circumference also doubles. If the diameter is 10 cm, then the circumference is 3.14 × 10 cm, or 31.40 cm. If the diameter is 20, then the circumference is 3.14 × 20 cm, or 62.80 cm.

b. The area is quadrupled. Explanations will vary. Sample student responses:

> You can draw circles on a grid. Each circle is about $\frac{3}{4}$ of the area of a square.
>
> The area of circle A is $\frac{3}{4} \times 2 \times 2 =$ 3 square units.
>
> The area of circle B is $\frac{3}{4} \times 4 \times 4 =$ 12 square units.

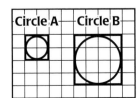

Circle A   Circle B

Overview Students read the Summary and solve three problems involving area, perimeter, volume, and circumference.

Planning Depending on students' ability to use the formulas for the circumference and area of a circle, you may want to discuss the way in which students are expected to use these formulas. Some students may find it helpful if the formulas are stated in words; for example, *the circumference is about three times the diameter,* and *the area is about three times the radius times the radius.* If students want to find more precise answers, they can use 3.14 instead of 3.

After the discussion, have students work individually on problems **29–31,** which can be used for assessment. Discuss these problems in class. After students complete this section, you may assign appropriate activities from the Try This! section, located on pages 49–52 of the Student Book, as homework.

Comments about the Problems

29–30. Informal Assessment These problems assess students' ability to understand the units and tools that are appropriate for estimating and measuring area, perimeter, volume, and circumference. They also assess students' ability to understand the structure and use of the metric and English systems of measurement.

31. Informal Assessment This problem assesses students' ability to analyze the effect of a systematic change in dimension on area and circumference.

If students have difficulty with this problem, suggest that they draw a circle and its enlargement. The circles may help students remember that doubling the circumference increases the area by a factor of four. Some students may be able to use formulas to find the circumference and the area. If not, encourage students to estimate these dimensions by using a strategy from this unit. Finally, allow students to use their estimates to answer the problem.

If students use the actual measurements of their drawings, ask them whether the relationship applies to other circles.

Assessment Overview

Students work on two assessment activities that you can use to collect additional information regarding students' understanding of area, perimeter, volume, and surface area and their ability to find the area of irregularly-shaped objects using standard and nonstandard measuring units.

Goals

- identify, describe, and classify geometric figures
- compare the areas of shapes using a variety of strategies and measuring units
- estimate and compute the areas of geometric figures
- create and work with tessellation patterns
- understand which units and tools are appropriate to estimate and measure area, perimeter, and volume
- understand the structure and use of standard systems of measurement, both metric and English
- use the concepts of perimeter, area, and volume to solve realistic problems
- represent and solve problems using geometric models
- analyze the effect a systematic change in dimension has on area, perimeter, and volume

Assessment Opportunities

Reallotment Problems

Sizing up Islands
Reallotment Problems

Reallotment Problems

Reallotment Problems

Sizing up Islands
Reallotment Problems

Reallotment Problems

Sizing up Islands
Reallotment Problems

Reallotment Problems

Reallotment Problems

Pacing

When combined, the Sizing up Islands and the Reallotment Problems assessments will take approximately three to four 45-minute class sessions. You may also assign these assessments as homework. See the Planning Assessment section on the next page for further suggestions as to how you might use the assessment activities.

About the Mathematics

These two assessment activities evaluate the major goals of the *Reallotment* unit. Refer to the Goals and Assessment Opportunities sections above for information regarding the specific goals evaluated in each assessment activity. Students may use different strategies to solve each problem. Their choice of strategy may indicate their level of comprehension of the problem. Consider how well students' strategies address the problem, as well as how successful students are at applying their strategies in the problem solving process. Students are not expected to use any formal algorithms to solve these problems.

The Sizing up Islands assessment provides students with an opportunity to investigate the area of an island of their choice and express the area in nonstandard units. The Reallotment Problems assessment is an assortment of 10 questions that allow students to demonstrate their understanding of area, perimeter, and volume. Students find the areas of shapes using standard measuring units, and they compare areas to find the prices of different-shaped items.

Materials

- Assessments, pages 137–141 of the Teacher Guide (one of each per student)
- atlases or world maps, page 113 of the Teacher Guide (one per student)
- centimeter and inch rulers, pages 113 and 115 of the Teacher Guide, optional (one per student)
- tracing paper, pages 117, 119, and 121 of the Teacher Guide, optional (one sheet per student)
- scissors, pages 117, 119, and 121 of the Teacher Guide, optional (one pair per student)
- drawing paper, page 121 of the Teacher Guide, optional (one sheet per student)

Planning Assessment

You may want students to work on these assessment problems individually if you want to evaluate each student's understanding and abilities. Make sure you allow enough time for students to complete the problems. Students are free to solve each problem in their own ways. They may choose to use any of the strategies or models introduced and developed in this unit to solve problems that do not call for a specific strategy or model.

The Sizing up Islands assessment may be given any time after students finish Section B. Depending on the time available, you may ask students to learn more about their chosen island and present their findings orally or in writing.

Scoring

Answers are provided for all assessment problems. The method of scoring the problems depends on the types of questions in each assessment. Most questions require students to explain their reasoning or justify their answers. For these questions, the reasoning used by the students in solving the problems as well as the correctness of the answers should be considered as part of your grading scheme. A holistic scoring scheme can be used to evaluation an entire task. For example, after reviewing a student's work, you may assign a key word such as *emerging, developing, accomplishing,* or *exceeding* to describe their mathematical problem-solving, reasoning, and communication.

On other tasks, it may be more appropriate to assign point values for each response. Student progress toward the goals of the unit should also be considered. Descriptive statements that include details of a student's solution to an assessment activity can be recorded. These statements would provide insight into a student's progress toward a specific goal of the unit. Descriptive statements can be more informative than recording only a score and can be used to document student growth in mathematics over time.

SIZING UP ISLANDS

Use additional paper as needed.

Find an island on a map or in an atlas. Estimate the area of this island using two different measuring units. Write a paragraph explaining the method and measuring units you used to estimate the island's area.

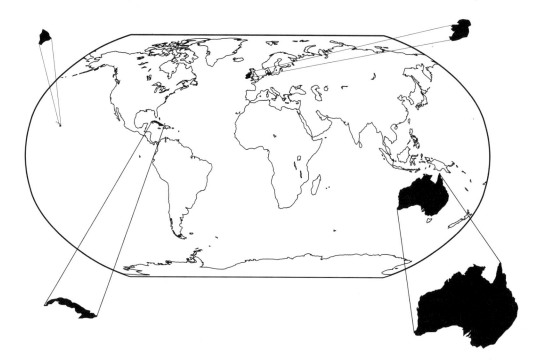

of student work

Answers will vary, depending on the island students choose. Sample student response:

I estimated that the area of Cuba is about 100,000 square kilometers. I used the scale on the map to draw a grid on tracing paper. Each square on the grid is about 10,000 square kilometers. Cuba covers a little more than 11 squares so its area is about 110,000 square kilometers. I also estimated that Cuba is about $\frac{3}{4}$ the size of Florida, because Florida covers about 15 squares on the grid.

Materials Sizing up Islands assessment, page 137 of the Teacher Guide (one per student); atlases or world maps (one per student); centimeter rulers, optional (one per student)

Overview Students choose an island from an atlas or world map and estimate its area using two different measuring units. They then write a paragraph that explains their choice of measuring units and method used to estimate the island's area.

About the Mathematics The purpose of this activity is to give students an opportunity to further develop their understanding of the concept of area: the number of identical units of measure that cover a surface. This problem assesses students' ability to estimate and compute area and to select appropriate units and tools to estimate and compute area.

Planning Encourage students to come up with their own methods of estimating the area of the island. This assessment can be used as a long range project anytime after students finish Section B or as an end-of-unit assessment. It may also be assigned as homework. You may want students to work on this activity in pairs or in small groups.

Extension If time permits, you might have each student compare the areas of two or three different islands or find the islands with the largest and smallest areas in the world.

Use additional paper as needed.

1. Find the area of each shape shown below, and explain how you found your answer.

a.

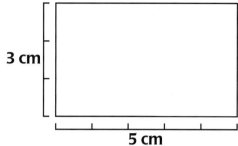

3 cm

5 cm

b.

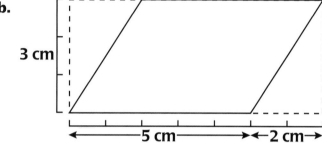

3 cm

5 cm   2 cm

c.

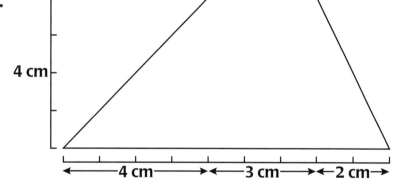

4 cm

4 cm   3 cm   2 cm

1. a. 15 square centimeters. Explanations will vary. Sample explanations:

There are three rows with five square centimeters in each row, so $3 \times 5 = 15$ square centimeters.

The area of a rectangle is $A = b \times h$. This rectangle has a base of 5 centimeters and a height of 3 centimeters so $A = 5 \times 3 = 15 \text{ cm}^2$.

b. 15 square centimeters. Explanations will vary. Sample explanations:

The area of a parallelogram is $A = b \times h$. This parallelogram has a base of 5 centimeters and a height of 3 centimeters so $A = 5 \times 3 = 15 \text{ cm}^2$.

I reshaped the parallelogram into a rectangle by cutting off a triangle from one side and pasting it onto the other side. The rectangle is the same size as the parallelogram. Since its area is 15 cm^2, the area of the parallelogram must also be 15 cm^2.

I found the area of the dotted rectangle: $(7 \times 3 = 21 \text{ cm}^2)$. Then I found the areas of the two triangles: $(2 \times \frac{1}{2} \times 2 \times 3 = 6 \text{ cm}^2)$. Then I subtracted the area of the triangles from the area of the dotted rectangle to find the area of the parallelogram: $21 \text{ cm}^2 - 6 \text{ cm}^2 = 15 \text{ cm}^2$.

c. 24 square centimeters. Explanations will vary. Sample explanations:

First I divided this shape into three parts and found the area of each part. Then I added the areas of the three parts to find the whole area.

Area of $a = \frac{1}{2}(4 \times 4) = 8 \text{ cm}^2$

Area of $b = 3 \times 4 = 12 \text{ cm}^2$

Area of $c = \frac{1}{2}(2 \times 4) = 4 \text{ cm}^2$

Total area $= 8 + 12 + 4 = 24 \text{ cm}^2$

Another strategy is to find the area of the rectangle enclosing the shape $(4 \times 9 = 36 \text{ cm}^2)$ and subtract the area of the triangle on the left $(\frac{1}{2} \times 4 \times 4 = 8 \text{ cm}^2)$ and the triangle on the right $(\frac{1}{2} \times 2 \times 4 = 4 \text{ cm}^2)$, so that $A = 36 - (8 + 4) = 24 \text{ cm}^2$.

Materials Reallotment Problems assessment, page 138 of the Teacher Guide (one per student); centimeter rulers, optional (one per student)

Overview Students find the areas of different geometric figures and explain the strategies they used to find their answers.

About the Mathematics This problem assesses students' ability to estimate and compute the areas of geometric figures. It also assesses their understanding of the units and tools that are appropriate to estimate and compute area, perimeter, and volume.

Planning Students may work on these assessment activities individually.

Comments about the Problems

1. a. Some students may draw squares in the rectangle and count the squares to find the area. Others may apply the formula for the area of a rectangle. Be sure that students show their work or give a written explanation that describes the strategies used to find the area.

b. Students may use one of the following strategies to compute the area:

- compute the area of the rectangles formed by the dotted lines and then subtract the areas of the two triangles outside of the parallelogram;

- reshape the parallelogram into a rectangle by cutting and pasting a triangular section; or

- use a formula to compute the parallelogram's area.

Some students may incorrectly multiply the parallelogram's base by the length of one of its sides (instead of the parallelogram's height) and therefore answer incorrectly.

c. Some students may divide the trapezoid into a rectangle and two triangles and find the area of each figure and add the results to find the total area. Others may enclose the entire figure within a rectangle, find the area of that rectangle, and subtract the areas of the unwanted regions.

Use additional paper as needed.

2. The grid on the right is made up of square centimeters. Find the area of the triangle. How did you find your answer?

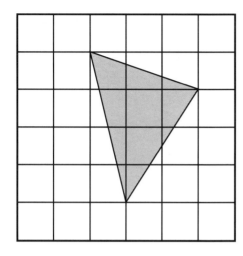

3. The square piece of chocolate shown below costs $1.00. Calculate the prices of the other two pieces of chocolate. Explain your answers. (*Note:* All the pieces have the same thickness.)

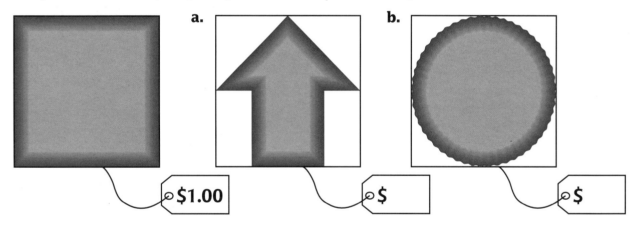

a.

b.

$1.00 $ $

c. Kirsten wants to buy the triangular piece of chocolate shown below that has been cut from the square piece above. How much will it cost? Explain your answer.

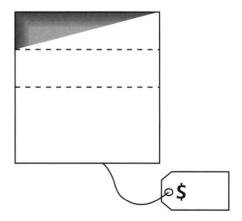

$

2. $5\frac{1}{2}$ square centimeters. Explanations will vary. Sample explanation:

First I drew a rectangle around the shaded triangle. The rectangle measures 3 cm by 4 cm, so its area is 3 cm × 4 cm = 12 cm². Then I found the areas of the three triangular spaces and added them together:

$A = \frac{1}{2} \times 1 \times 4 = 2$ cm²

$A = \frac{1}{2} \times 3 \times 1 = 1\frac{1}{2}$ cm²

$A = \frac{1}{2} \times 3 \times 2 = 3$ cm²

Then I subtracted the area of the triangular spaces from the area of the rectangle to find the area of shaded triangle.

$A = 12$ cm² $- 6\frac{1}{2}$ cm² $= 5\frac{1}{2}$ cm²

3. a. 50¢. Explanations will vary. Sample explanation:

I cut the arrow in half lengthwise, and then I cut those pieces in half crosswise. I pasted the two rectangles together to make one square and I pasted the two triangles together to make another square. Each of these small squares is $\frac{1}{4}$ of the original square. The area is $\frac{1}{4} + \frac{1}{4} = \frac{1}{2}$ of the square. The cost of this piece is $\frac{1}{2} \times 100¢ = 50¢$.

b. 75¢. Explanations will vary. Sample explanation:

The chocolate covers a little more than $\frac{3}{4}$ of the square. The cost is $\frac{3}{4} \times 100¢ = 75¢$.

c. About 12 or 13¢. Explanations will vary. Sample explanation:

The chocolate covers $\frac{1}{2}$ of $\frac{1}{4}$ of the square. The size of this piece is $\frac{1}{2} \times \frac{1}{4} = \frac{1}{8}$. Cost $= \frac{1}{8} \times 100¢ =$ about 12 or 13¢

Materials Reallotment Problems assessment, page 139 of the Teacher Guide (one per student); tracing paper, optional (one sheet per student); scissors, optional (one pair per student)

Overview Students find the area of a triangle. They also determine the relative cost of different-shaped pieces of chocolate, given the area and cost of one square piece.

About the Mathematics Problem **2** assesses students' ability to compare the areas of shapes using a variety of strategies and measuring units. Problem **3** assesses students' ability to:

• estimate, measure, and compute the areas of geometric figures,

• select appropriate units and tools to estimate and measure area, and

• use the concept of area to solve realistic problems.

Planning Students may work on these assessment activities individually.

Comments about the Problems

2. Since the base and height of the triangle are unknown, students cannot use a formula here. If students have difficulty, suggest that they draw a rectangle around the triangle, find the area of the rectangle, and subtract from it the areas of the unshaded triangles.

3. Be sure to discuss students' solutions and strategies they used to determine the cost of each piece of chocolate.

a. Students may use tracing paper and scissors to reallot parts of the shaded region to estimate and compare its area with that of the given square region.

b. Students may estimate the area of the circular shape and determine that it is more than half of the area of the square. Other students may use the formula for the area of a circle. Again, give students the option of using tracing paper and scissors.

c. Some students may need to draw a line to divide the lower half of the square into two equal parts, making a total of four equal parts. Other students may observe that the area of the shaded part makes up $\frac{1}{2}$ of $\frac{1}{4}$, or $\frac{1}{8}$, of the square.

Use additional paper as needed.

4. The piece of wood shown below measures two meters by three meters and costs $18. Find the price of the other three pieces of wood. Explain your answer. (*Note:* All the pieces of wood have the same thickness.)

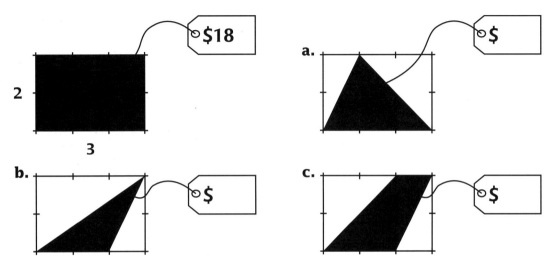

5. Tanya wants to plant grass seed in a small yard that measures six meters long and four meters wide. The clerk at Handy Dandy's Garden Shop gives her this order form to fill out. Complete the form showing how many bags of each size of grass seed are needed and the total cost. Explain your answers.

Handy Dandy's Garden Shop • Lawn Seed Ordering Form				
Size of Grass Seed Bag	Maximum Square Footage Covered	Cost	Total Bags Ordered	Total Cost
Small	4 sq m²	$3.75		
Medium	8 sq m²	$7.25		
Large	10 sq m²	$8.95		
Extra Large	20 sq m²	$17.50		

6. Look at the tessellation below. Find the area of one bird. Explain how you found the answer.

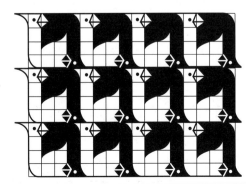

4. a. $9. Explanations will vary. Sample explanation:

The base and height of the triangle are the same as the length and width of the rectangle. The area of the triangle is half the area of the rectangle, so the triangle costs $\frac{1}{2} \times \$18 = \$9$.

b. $6. Explanations will vary. Sample explanations:

The price per square meter is $\$18 \div 6 = \$3$. The area of the triangular piece of wood is $\frac{1}{2} \times 2\text{ m} \times 2\text{ m} = 2\text{ m}^2$. The cost is $2 \times \$3 = \$6$.

I subtracted the cost of the two triangles from the cost of the rectangle. The area of the triangle on the left is $\frac{1}{2} \times 3 \times 2 = 3\text{ m}^2$. The area of the triangle on the right is $\frac{1}{2} \times 2 \times 1 = 1\text{ m}^2$. The area of the rectangle is $2 \times 3 = 6\text{ m}^2$. So the area of the remaining triangle is $6\text{ m}^2 - 4\text{ m}^2 = 2\text{ m}^2$. The cost is $2 \times \$3 = \$6$.

c. $9. Explanations will vary. Sample explanation:

The piece of wood can be divided into two triangles. One triangle has an area of 2 m^2 ($\frac{1}{2} \times 2 \times 2$). The other triangle has an area of 1 m^2 ($\frac{1}{2} \times 2 \times 1$). The total area of the shape is 3 m^2, so the cost is $3 \times \$3$ or $9.

5. Answers and explanations will vary. Sample response:

The lawn is $6 \times 4 = 24\text{ m}^2$. Tanya plans to buy six small bags of grass. Since each small bag covers 4 square meters, 6 bags will cover 24 square meters. Six bags will cost $6 \times \$3.75 = \$22.50$.

6. 8 square units. Explanations will vary. Sample explanation:

Two birds cover an area of 16 square units, so one bird covers half of that or 8 square units.

There are $16 \times 12 = 192$ square units in the tessellation. There are 24 birds (12 white and 12 black birds). There are $192 \div 24 = 8$ square units per bird.

Materials Reallotment Problems assessment, page 140 of the Teacher Guide (one per student); tracing paper, optional (one sheet per student); scissors, optional (one pair per student)

Overview Given the dimensions and cost of one piece of wood, students estimate the relative costs of three other pieces of wood. They also find the area of a land plot to determine the quantity and cost of grass seed needed to cover it. Students then examine a tessellation and find the area of one figure in the repeating pattern.

About the Mathematics Problems **4** and **6** assess students' ability to estimate and compute the areas of geometric figures, select appropriate units and tools to estimate and measure area, and use the concept of area to solve realistic problems. Problems **4** and **5** assesses students' ability to solve problems using geometric models.

Planning Students may work on these assessment activities individually.

Comments about the Problems

4. Students may use one of the following strategies to find the area of the shaded region:

   • count the number of whole squares in the shaded region and estimate the number of squares that the remaining pieces will make, or

   • find the total area of the rectangle that encloses the shaded region and subtract the areas of unshaded parts from the total area of the rectangle.

5. Some students may need to sketch the rectangular lawn to help them find the area. Others may multiply the two given dimensions to find the area.

6. Reallotting small parts of one bird figure in the tessellation to make whole squares is very difficult. If students are having difficulty, you might encourage them to count the number of squares in a relatively large portion of the tessellation and divide that by the number of birds within that same area to estimate the area of one bird.

Extension: As an extension to problem **5,** you may ask students whether the cost of grass seed for one square meter decreases as the package size increases.

Use additional paper as needed.

7. Rocky the cat is sleeping on a handmade quilt. How many small triangular pieces were used to make the quilt? Explain your answer.

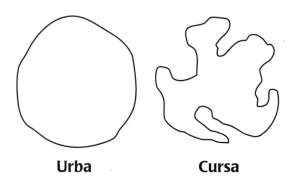

Urba   **Cursa**

8. Shown on the left are maps of two islands, Urba and Cursa. Which one has the larger area? Explain your answer.

9. Emile used a parallelogram-shaped piece of paper to make the cylinder shown below. He wants to make a cylinder with a diameter that is twice as large as this one. What size parallelogram does he need? Explain your answer.

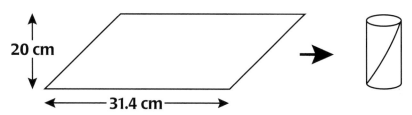

20 cm

◄—— 31.4 cm ——►

10. A jigsaw puzzle comes in a box (shown below) that is 25 centimeters by 35 centimeters by 4 centimeters.

a. You need cartons to ship these boxed puzzles to a toy store. Design a carton that holds exactly 30 of these jigsaw puzzle boxes. Explain your choice.

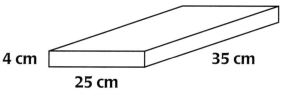

4 cm   **35 cm**

25 cm

b. Do you think all possible jigsaw puzzle box designs will use the same amount of cardboard? Explain your answer.

7. 48 triangular pieces. Explanations will vary. Sample explanation:

There are eight hexagon shapes in the quilt. Each hexagon is made of 6 triangles. $8 \times 6 = 48$ triangular pieces.

8. Urba has the larger area. Explanations will vary. Sample explanation:

If you trace and cut out the shapes for both islands and place the Cursa shape over the Urba shape, a lot of the shape for Urba remains uncovered.

9. Base = 62.8 centimeters and height = 20 centimeters. Explanations will vary. Sample explanations:

The base of this parallelogram represents the circumference of the cylinder. In the original cylinder, the circumference is 31.4 centimeters. To find the diameter, divide by π or 3.14. The diameter of the original cylinder is 10 centimeters. The diameter of the new cylinder is 2×10 cm = 20 cm. The circumference of the new cylinder is 20 cm $\times$ 3.14 = 6.28 cm. The height of the cylinder does not change. It is still 20 cm.

The formula for circumference = $\pi \times$ d. If d (the diameter) doubles, the circumference also doubles. The circumference is the same as the base of this parallelogram. So the circumference of the new cylinder is 2×31.4 cm = 62.8 cm. The height stays the same.

10. **a.** Designs will vary. Sample student response:

My carton measures 35 centimeters by 50 centimeters by 60 centimeters. Two puzzle boxes fit side by side in one layer, which measures 35 centimeters by (25 + 25) centimeters by 4 centimeters. Because there are two puzzles in each layer, the box needs to hold $30 \div 2 = 15$ layers. Since each layer is 4 centimeters tall, the box is 4 cm $\times$ 15 = 60 cm tall.

Other possible designs for rectangular boxes include:

Length	Width	Height
35 cm	25 cm	120 cm
70 cm	25 cm	60 cm
105 cm	25 cm	40 cm
35 cm	75 cm	40 cm

b. No. Explanations will vary. Sample explanation:

Different designs will use different amounts of cardboard. A carton measuring 35 cm by 25 cm by 120 cm will take 16,150 square cm of cardboard to make. A carton measuring 70 cm by 25 cm by 60 cm needs 14,900 square cm of cardboard.

Materials Reallotment Problems assessment, page 141 of the Teacher Guide (one per student); drawing paper, optional (one sheet per student); scissors, optional (one pair per student)

Overview Students use counting strategies to find the total number of triangular pieces in a quilt. They also compare the areas of two islands. Given the dimensions of a parallelogram-shaped piece of paper used to make a cylinder, students describe the dimensions needed to make a cylinder twice as large.

About the Mathematics Problems **7** and **8** assess students' ability to select appropriate units and tools to estimate and measure area. Problem **9** assesses students' ability to use the concepts of perimeter and volume to solve realistic problems; analyze the effect a systematic change in dimension has on perimeter and volume; and generalize formulas and procedures for determining the circumference of a circle.

Planning Students may work on these assessment activities individually.

Comments about the Problems

7. Encourage students to find a pattern or use a counting strategy instead of counting each triangular piece.

8. Students' responses to this problem will reveal whether or not they incorrectly think that the shape with the larger perimeter (the island of Cursa) also has the larger area. Most students should be able to visually determine that Urba has the larger area. Give students the option of using tracing paper and scissors to cut out the two shapes and lay one shape on top of the other to compare the areas.

9. Some students may want to physically model this problem by drawing the parallelogram, cutting it out, and forming it to make a cylinder.

10. **a.** Encourage students to use their own strategies to design a shipping carton. If students are having difficulty, suggest that they find the length of two puzzle boxes lying side by side (70 cm) and use that answer to determine the length of the shipping carton. Ask students to find the width and height of the shipping carton using the same strategy.

Reallotment
Glossary

The Glossary defines all vocabulary words listed on the Section Opener pages. It includes the mathematical terms that may be new to students, as well as words having to do with the contexts introduced in the unit. (*Note:* The Student Book has no glossary. This is in order to allow students to construct their own definitions, based on their personal experiences with the unit activities.)

acre (p. 32) a unit of area commonly used in measuring land in the United States; one acre is equal to 43,560 square feet

area (p. 14) the number of measuring units needed to cover a shape

base (p. 50) the width of the bottom of a figure

circumference (p. 94) the perimeter of a circle

contiguous (p. 28) physically connected (as in the 48 contiguous states of the United States)

diagonal (p. 48) a line segment connecting two nonadjacent corners of a figure

diameter (p. 94) the length of a line segment through the center of a circle or the actual physical segment

height (p. 50) the vertical distance from the bottom to the top of something

hexagon (p. 94) a six-sided polygon

kilometer (p. 68) a metric unit of length equal to 1,000 meters and approximately equal to 0.62 mile

liter (p. 100) a metric unit of volume equal to 1,000 cubic centimeters or the amount of liquid a 10-centimeter cube holds

measuring unit (p. 14) a specific amount used to measure size; measuring units may be standard, such as an inch or a centimeter, or they may be nonstandard, such as the length of a paper clip

meter (p. 46) a metric unit of length equal to 100 centimeters and approximately equal to 39 inches

parallelogram (p. 42) a quadrilateral with opposite sides parallel

perimeter (p. 86) the distance around a shape

polygon (p. 17) a closed plane figure bounded by straight line segments

population density (p. 82) the number of members per unit of area

quadrilateral (p. 42) a four-sided polygon

radius (p. 96) the length of a line segment between the center of a circle and its circumference

reallotment (p. 17) the reshaping of a figure to form a new figure in which the area is not changed

regular (p. 94) having equal sides and equal angles

reshape (p. 15) to transform a shape by cutting and pasting

surface area (p. 98) the number of measuring units needed to cover the exterior region of an object

tessellation (p. 16) a repeating pattern with no open spaces that completely covers a region

volume (p. 98) the amount of space that a three-dimensional figure occupies

Blackline Masters

Dear Family,

Your child is about to begin the *Mathematics in Context* unit *Reallotment*. Below is a letter to your child describing the unit and its goals.

This unit introduces the term *reallotment* which means the reshaping of a figure to form a new figure with the same area as the original. Reallotment often makes it easier to find and compare the areas of different shapes.

You can help your child relate the class work to his or her own life through a variety of at-home activities. You might ask your child to figure out a way to compare the areas of your living room and dining room to see which has the larger area.

You might also ask your child to estimate the cost of installing a fence to enclose your property or to estimate the cost of installing new carpet in the family room.

Students are introduced to tessellations, as shown in the bottom picture of the Student Letter. You might encourage your child to look for tessellation patterns in your home, neighborhood, or local area. You could also ask your child to draw some creative tessellation designs and explain what the repeating pattern is in each design.

Dear Student,

Welcome to the unit *Reallotment*.

In this unit you will study different shapes and the space covered by a variety of shapes.

You will figure out things such as how many people can stand in your classroom. How could you find out without packing people in the entire classroom?

You will also investigate the border around a shape and the amount of space inside a three-dimensional figure.

How can you make a shape like the one below that will cover a floor, leaving no open spaces?

In the end, you will have learned some important ideas about algebra, geometry, and arithmetic. We hope you enjoy the unit.

Sincerely,

The Mathematics in Context Development Team

Have fun helping your child make these connections between mathematics and the real world!

Sincerely,

The Mathematics in Context Development Team

Name_____

3. Which field has the most tulip plants? Justify your answer.

Field A 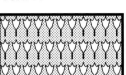   **Field B**

Field C

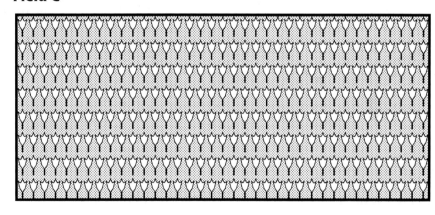

Use with *Reallotment,* page 3.

Mary Ann decides that 80¢ is a reasonable price for the big square piece (figure **a** below).

4. What should the other prices be? (*Note:* All of the pieces have the same thickness.)

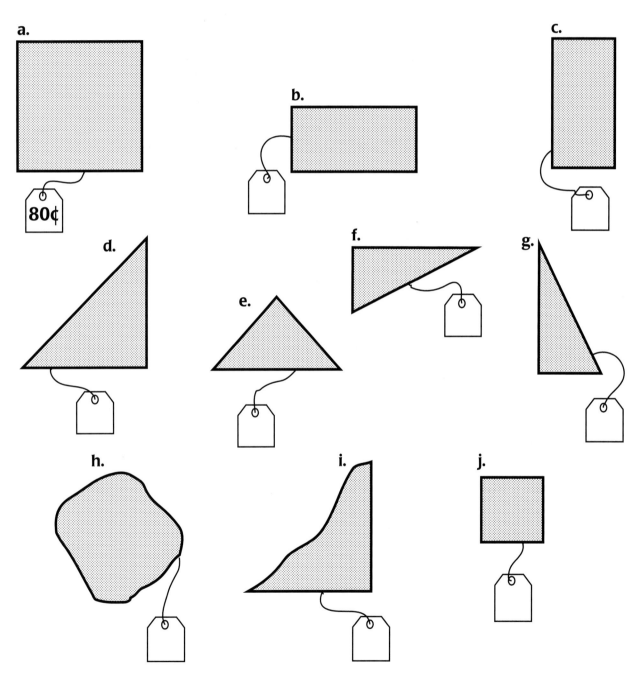

Name_____

The small square tile (**a**) costs $5.

6. Figure out fair prices for the other tiles. Discuss your strategies with some of your classmates.

Name _____

Use with *Reallotment,* pages 6 and 37.

Name_____

4. Compare the area of your state to the area of the 48 contiguous United States.

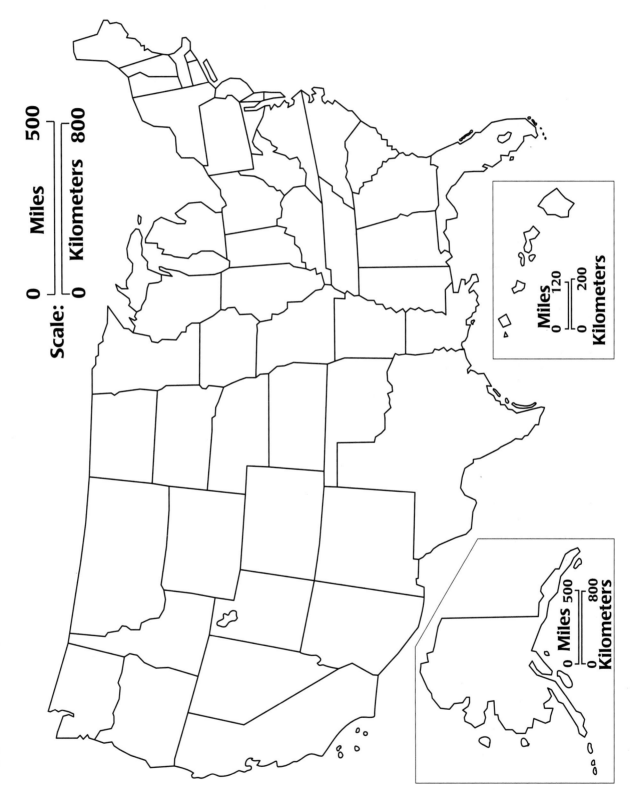

Use with *Reallotment,* page 12.

5. **a.** Which island is bigger? How do you know?

   b. Estimate the area of each island.

Smoke Island

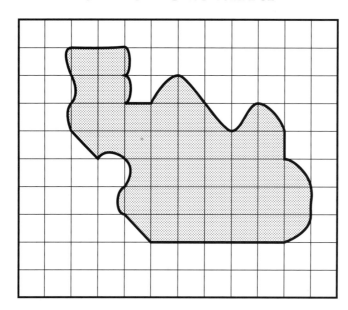

Fish Island

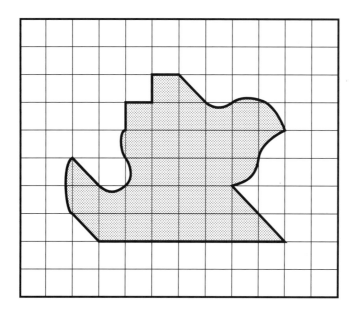

10. Determine the area of each of the shaded pieces below. Give your answers in square units. Be prepared to explain your reasoning.

Use with *Reallotment,* page 14.

Name _____

11. Calculate the prices of the shaded pieces.

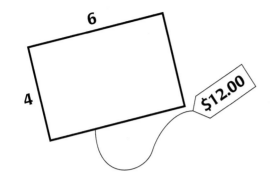

$12.00

a.
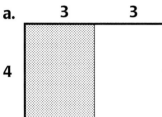

b.
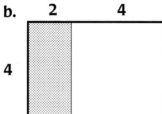

c.
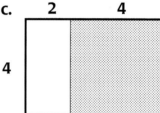

d.
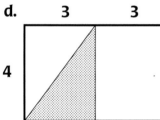

e.
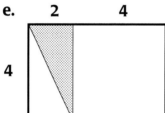

f.
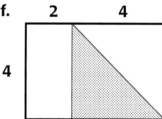

g.
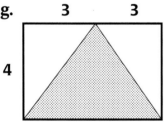

h.
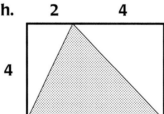

i.
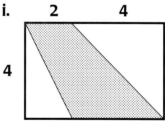

j.
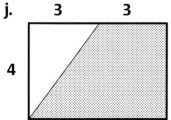

k.
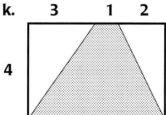

l.
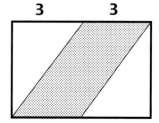

17. How many small tiles were used to create the floor? (Think of a fast way to count the tiles.)

A Small Tile

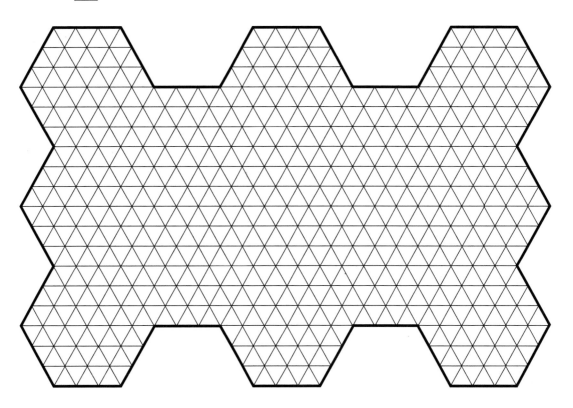

Use with *Reallotment,* page 24.

Name _____

**A Large Section
of Tile**

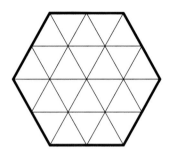

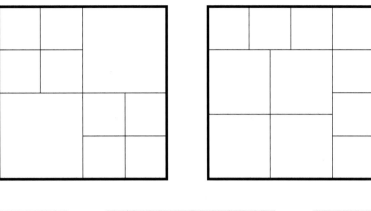

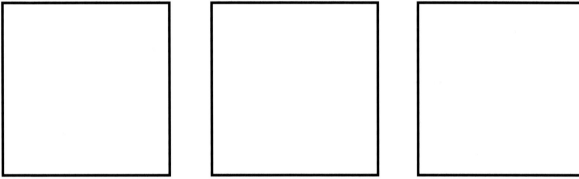

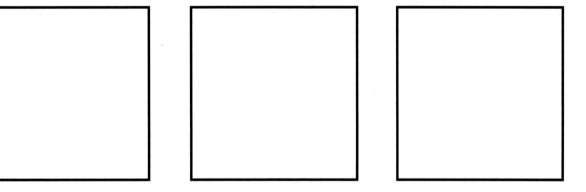

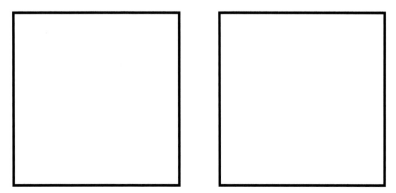

Use with *Reallotment*, page 42.

18. Cut apart the sections of the circle below. Rearrange them to help you develop a formula for the area of a circle.

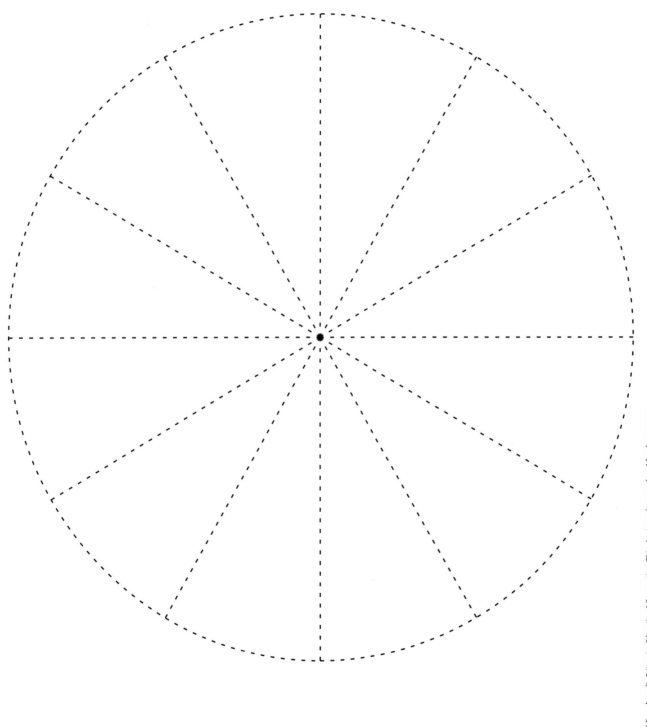

SIZING UP ISLANDS

Use additional paper as needed.

Find an island on a map or in an atlas. Estimate the area of this island using two different measuring units. Write a paragraph explaining the method and measuring units you used to estimate the island's area.

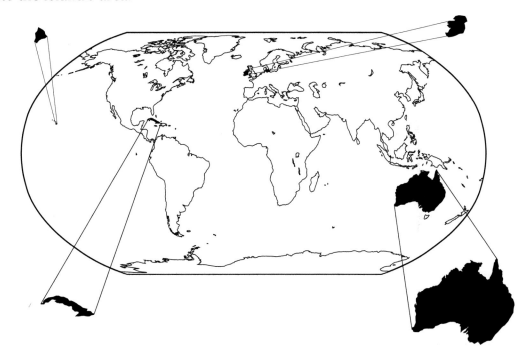

REALLOTMENT PROBLEMS

Use additional paper as needed.

1. Find the area of each shape shown below, and explain how you found your answer.

a.

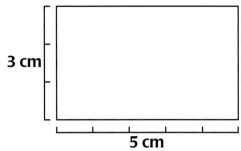

b.

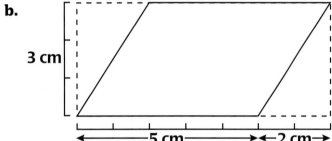

c.

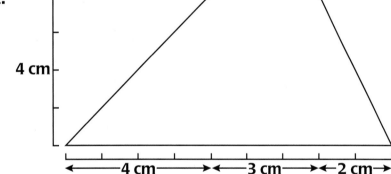

Use additional paper as needed.

2. The grid on the right is made up of square centimeters. Find the area of the triangle. How did you find your answer?

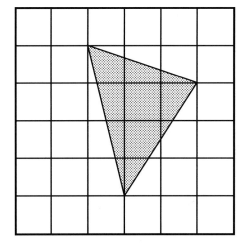

3. The square piece of chocolate shown below costs $1.00. Calculate the prices of the other two pieces of chocolate. Explain your answers. (*Note:* All the pieces have the same thickness.)

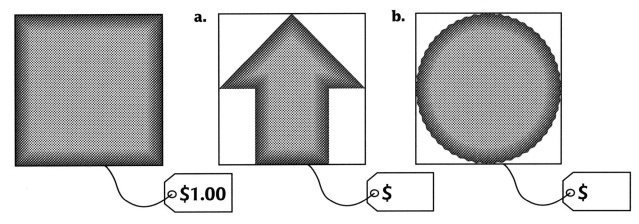

a.

b.

$1.00 $ $

c. Kirsten wants to buy the triangular piece of chocolate shown below that has been cut from the square piece above. How much will it cost? Explain your answer.

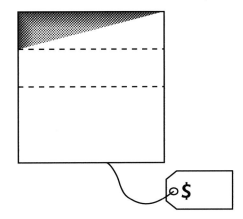

$

REALLOTMENT PROBLEMS

Use additional paper as needed.

4. The piece of wood shown below measures two meters by three meters and costs $18. Find the price of the other three pieces of wood. Explain your answer. (*Note:* All the pieces of wood have the same thickness.)

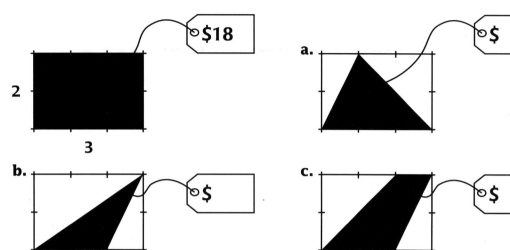

5. Tanya wants to plant grass seed in a small yard that measures six meters long and four meters wide. The clerk at Handy Dandy's Garden Shop gives her this order form to fill out. Complete the form showing how many bags of each size of grass seed are needed and the total cost. Explain your answers.

Handy Dandy's Garden Shop • Lawn Seed Ordering Form				
Size of Grass Seed Bag	Maximum Square Footage Covered	Cost	Total Bags Ordered	Total Cost
Small	4 sq m²	$3.75		
Medium	8 sq m²	$7.25		
Large	10 sq m²	$8.95		
Extra Large	20 sq m²	$17.50		

6. Look at the tessellation below. Find the area of one bird. Explain how you found the answer.

Use additional paper as needed.

7. Rocky the cat is sleeping on a handmade quilt. How many small triangular pieces were used to make the quilt? Explain your answer.

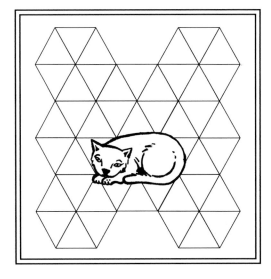

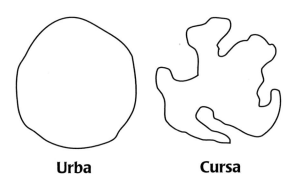

Urba   **Cursa**

8. Shown on the left are maps of two islands, Urba and Cursa. Which one has the larger area? Explain your answer.

9. Emile used a parallelogram-shaped piece of paper to make the cylinder shown below. He wants to make a cylinder with a diameter that is twice as large as this one. What size parallelogram does he need? Explain your answer.

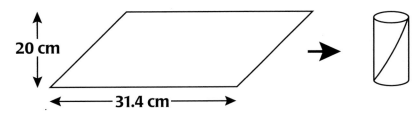

20 cm

31.4 cm

10. A jigsaw puzzle comes in a box (shown below) that is 25 centimeters by 35 centimeters by 4 centimeters.

a. You need cartons to ship these boxed puzzles to a toy store. Design a carton that holds exactly 30 of these jigsaw puzzle boxes. Explain your choice.

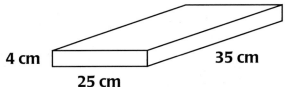

4 cm   35 cm

25 cm

b. Do you think all possible jigsaw puzzle box designs will use the same amount of cardboard? Explain your answer.

Section A. The Size of Shapes

1. **a.** $5

    b. $2.50

    c. $5

    d. $2.50

    e. $3.75

    f. $1.25

    Strategies will vary. Sample strategy:

    The area of part **a** is $\frac{1}{4}$ of the total area of the board, so part **a** will cost $\frac{1}{4}$ of $20, or $5.

    The area of part **b** is $\frac{1}{2}$ of the area of part **a,** so part **b** will cost $2.50.

    The area of part **c** is twice as large as that of part **b,** so part **c** will cost $5;

    The area of part **d** is $\frac{1}{2}$ of the area of part **a,** so part **d** will cost $2.50.

    The area of part **f** is $\frac{1}{2}$ of the area of part **b,** so part **f** will cost $1.25.

    The area of part **e** is equal to the areas of parts **d** and **f,** so part **e** will cost $2.50 + $1.25 = $3.75.

2. **a.** The area is four square units. Strategies will vary. Sample strategy:

    I added together the number of whole squares (1) and half squares (6) in the shape:
    $1 + 6 \times \frac{1}{2} = 1 + 3 = 4$ square units.

    b. Sample tessellation:

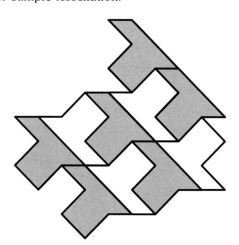

Section B. Areas

1. **a.** $6

    b. $3.75

    c. $11.25

    d. $6

    e. $11.25

    f. $6

    Strategies will vary. Sample strategy for part **a:**

    I found the area of the rectangle by counting the total number of squares in the shape. It has an area of 8 square units. The area of the shaded region (the fiberglass) is one-half the area of the rectangle, so its area is 4 square units. Each piece of fiberglass that has an area of 4 square units costs $6, so part a costs $6.

2. Drawings will vary. Sample drawings:

    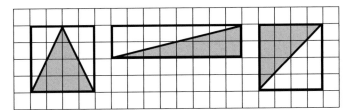

Section C. Area Patterns

1. Drawings will vary. Sample drawing:

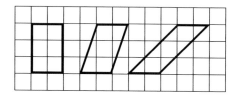

   Each parallelogram has a base of 2 units and a height of 3 units.

2. **a.** six square units

   b. four square units

   c. 1.5 square units

   Strategies will vary. Sample strategy for finding the area of part **c:**

   I traced the triangle on another sheet of paper and reshaped the triangle by cutting and pasting parts that would fill the square units on the grid. The triangle has an area of about 1.5 square units.

3. Answers will vary. Sample responses:

   Some students may trace the kitchen floor on another sheet of paper and divide it into large and medium sections of tile, as shown below:

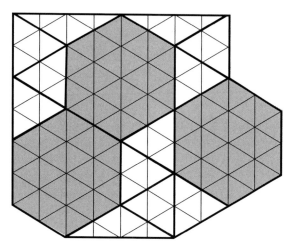

Kitchen Floor

   • The kitchen floor (shown above) is made up of 3 large sections and 12 medium sections of tile (three cut in half vertically).

   • The kitchen floor is made up of 3 large sections, 9 medium sections (two cut in half vertically), and 12 individual tiles (two cut in half vertically).

   • The kitchen floor is made up of 3 large sections, 10 medium sections (three cut in half vertically), and 8 individual tiles.

Section D. Measuring Area

1. a. Drawings will vary. Sample drawing:

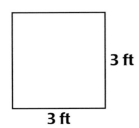

3 ft

3 ft

   b. Each side of the square is 1 yard. The area of the square is 1 square yard.

   c. Each side of the square is 36 inches. The area of the square is 1,296 square inches.

2. a. 5 square meters = 50,000 square centimeters. Strategies will vary. Sample strategy:

First I drew a rectangle that has an area of 5 square meters and labeled the length and width of one square meter to show its measurements, as shown below.

5 square meters

1 m | 10,000 cm²

1 m

Then I converted the measurement for each side from meters to centimeters.
Since 1 meter = 100 centimeters, each side measures 100 centimeters. So, the area of each square meter is 100 cm × 100 cm, or 10,000 square centimeters. So, the area of 5 square meters = 50,000 square centimeters.

   b. 3 square feet = 432 square inches. Strategies will vary. Sample strategy.

First I drew a rectangle that has an area of 3 square feet and labeled the length and width of one square foot to show its measurements, as shown below.

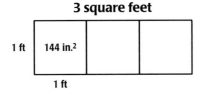

3 square feet

1 ft | 144 in.²

1 ft

Then I converted the measurement for each side from feet to inches. Since 1 foot = 12 inches, each side measures 12 inches. So, the area of one square foot is 12 in. × 12 in. or 144 square inches. So, the total area of 3 square feet = 144 × 3 = 432 square inches.

c. 18 square feet = 2 square yards. Strategies will vary. Sample strategy.

First I drew a rectangle that has an area of 18 square feet. Then I divided the rectangle into two squares and labeled the length and width of each square to show their measurements, as shown below.

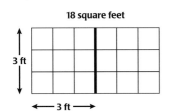

Then I converted the measurement for the side of each square from feet to yards. Since 3 feet = 1 yard, each side measures 1 yard. So, the area of each square is 1 yd × 1 yd, or 1 square yard and the area of the rectangle is 2 square yards.

d. 50 square centimeters = 5,000 square millimeters. Strategies will vary. Sample strategy:

First I drew a square that has an area of 25 square centimeters and labeled the length and width of the square to show its measurements, as shown below.

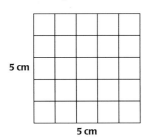

Then I converted the measurement for its length and width from centimeters to millimeters. Since 1 centimeter = 10 millimeters, each side of the square measures 50 millimeters. So, the area of one square = 50 × 50 = 2,500 square millimeters. The total area of two squares is 2,500 × 2 = 5,000 square millimeters.

3. Reports will vary. Students should include the following information in their reports:

total area of office: 4 m × 5.25 m = 21 square meters
total cost of carpet: $252
total cost of large tiles: $273
total cost of small tiles: $336

4. a. 25,000,000 square meters. Strategies will vary. Sample strategy:

I calculated 250 million ÷ 10 = 25 million square meters.

b. 25 square kilometers. Strategies will vary. Sample strategy:

There are 1,000 meters in one kilometer, and 1,000 m × 1,000 m, or 1,000,000 square meters in one square kilometer. So, there are 25 square kilometers in 25 million square meters.

c. Answers will vary, depending on the state or city students are familiar with. For example, the area of Rhode Island is 3,144 km², and the area of Chicago is 30 km².

Section E. Perimeter, Area, and Volume

1. **a.** Perimeter = 10 units; Area = 6 square units.

   b. Perimeter = 14 units; Area = 6 square units.

   c. Perimeter = 11 units; Area = 6 square units.

   d. Perimeter = 25 units; Area = 6 square units.

   The area of each rectangle is the same, but each rectangle has a different perimeter.

2. Tables will vary. Sample table:

Length (in cm)	Width (in cm)	Height (in cm)	Volume (in cm³)	Surface Area (in cm²)
20	1	1	20	82
5	4	1	20	58
5	2	2	20	48
10	2	1	20	64

3. **a.**

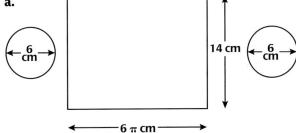

   b. Area of top: 9π or 28.3 square centimeters

   c. Circumference of container: 6π or 18.8 square centimeters

   d. The surface area is approximately 320 square centimeters. Explanations will vary. Students should add the areas of the top, bottom, and side of the container. The top and the bottom have the same area (28.3 cm²). The area of the side of the container is equal to its circumference (6π cm) times its height (14 cm), which is equal to about 263.2 cm². So, the total surface area = (2 × 28.3 cm²) + 263.2 cm² = 319.8 cm².

CREDITS

Cover

Design by Ralph Paquet/Encyclopædia Britannica Educational Corporation.

Collage by Koorosh Jamalpur/KJ Graphics.

Illustrations

12, 42, 90, 92, 100 Phil Geib/Encyclopædia Britannica Educational Corporation.

Photographs

80 © Russ Hansen Films; **104** © Ralph Arvidson. Reprinted with permission, *Chicago Sun-Times* © 1996.

Special Thanks

Architectural drawings on page 96 were made by Reiner A. Pligge, President of Reiner A. Pligge Architects in Madison, WI.

Mathematics in Context is a registered trademark of Encyclopædia Britannica Educational Corporation. Other trademarks are registered trademarks of their respective owners.